Gene Expression and Its Discontents

Rodrick Wallace · Deborah Wallace

Gene Expression and Its Discontents

The Social Production of Chronic Disease

Rodrick Wallace
New York State Psychiatric Institute
Division of Epidemiology
1051 Riverside Dr.
New York NY 10032
USA
wallace@pi.cpmc.columbia.edu

Deborah Wallace
549 West 123rd Street
New York NY 10032
Apt. 16F
USA
rdwall@ix.netcom.com

ISBN 978-1-4419-1481-1   e-ISBN 978-1-4419-1482-8
DOI 10.1007/978-1-4419-1482-8
Springer New York Dordrecht Heidelberg London

Library of Congress Control Number: 2009938266

© Springer Science+Business Media, LLC 2010
All rights reserved. This work may not be translated or copied in whole or in part without the written permission of the publisher (Springer Science+Business Media, LLC, 233 Spring Street, New York, NY 10013, USA), except for brief excerpts in connection with reviews or scholarly analysis. Use in connection with any form of information storage and retrieval, electronic adaptation, computer software, or by similar or dissimilar methodology now known or hereafter developed is forbidden.
The use in this publication of trade names, trademarks, service marks, and similar terms, even if they are not identified as such, is not to be taken as an expression of opinion as to whether or not they are subject to proprietary rights.

Printed on acid-free paper

Springer is part of Springer Science+Business Media (www.springer.com)

The first six chapters of this book use the asymptotic limit theorems of information theory to understand how epigenetic context affects organismal development by invoking a cognitive paradigm for gene expression. A simple argument suggests that epigenetic information sources act as analogs to a tunable catalyst, directing development to different characteristic pathways in a manner similar to ecosystem resilience shifts. The results have significant implications for epigenetic epidemiology, showing how environmental stressors, in a large sense, can induce a spectrum of chronic disorders. Chapters 7-11 apply the perspective to a number of diseases broadly associated with obesity that are becoming pandemic, using US data at different scales of observation. The theory appears to work very well at individual and simple aggregate levels for a number of chronic conditions in stressed populations. Environments that can be characterized as having regularities of 'grammar' and 'syntax' can interact with organismal development via epigenetic catalysis to literally write distorted images of themselves onto the human life course in a highly plieotropic and often punctuated manner, producing trajectories to a range of serious dysfunctions.

Chapter 12, however, examines the appearance of a 'chronic disease guild' at the neighborhood level, suggesting that culturally structured community stressors may lock-in epigenetic effects, a synergism across three systems of human heritage that will require further study to explicate. Communities, in which individuals respond collectively, thus display complicated patterns of chronic disease that may represent another example of a 'mesoscale resonance' in which the dynamics of ecological keystone structures entrain phenomena at other scales.

One implication of this work is that pandemic chronic diseases at both the individual, via plieotropy, and the population scales, via collective modalities of gene expression, are unlikely to respond to individual-level – i.e., medical – interventions in the face of serious, persistent individual and community stress, and may require large scale changes in public policy and resource allocation for their amelioration. Drugs powerful enough to affect deleterious epigenetic programming are likely to trigger profound iatrogenic 'side effects' that, over the life course, would not only obviate the intervention, but most likely lead to shortened lifespans.

The book, a synthesis across half a dozen peer-reviewed publications, can be read at several levels. Chapters 1 and 6-13 form a natural introductory unit that can be followed by the more mathematical sections as desired.

Contents

1 Introduction 1
   1.1 Toward new tools 1
   1.2 Epigenetic epidemiology 2
   1.3 Ecosystem resilience 6
   1.4 A first survey of the obesity pandemic in the US 15

2 Models of development 27
   2.1 The spinglass model 27
   2.2 Shifting perspective: cognition as an information source 30

3 Groupoid symmetries 37
   3.1 The first level 38
   3.2 The second level 38
   3.3 Spontaneous symmetry breaking 40
   3.4 A biological example 41

4 Epigenetic catalysis 45
   4.1 The basic idea 45
   4.2 Rate Distortion dynamics 47
   4.3 More topology 50
   4.4 Inherited epigenetic memory 51
   4.5 Multiple processes 52
   4.6 'Coevolutionary' development 53
   4.7 Multiple models 55
   4.8 Epigenetic focus 56

5 Developmental disorders 59
   5.1 Network information theory 59
   5.2 Embedding ecosystems as information sources 60
   5.3 Ecosystems farm organismal development 61
   5.4 A simple probability argument 64

		5.5 Developmental shadows 65

      5.5 Developmental shadows 65
      5.6 Epigenetic programming of artificial systems for biotechnology 66

6 An interim perspective 69

7 The obesity pandemic in the US 73
      7.1 Introduction 73
      7.2 Stress and the HPA axis..................................... 74
      7.3 HPA axis cognition 75
      7.4 Interacting information sources............................. 75
      7.5 The simplest HPA axis model 77
      7.6 Obesity as a developmental disorder 81
      7.7 Recent trajectories of structured stress in the US 82
      7.8 Confronting the obesity epidemic 89

8 Coronary heart disease in the US 99
      8.1 Introduction 99
      8.2 Cognition, immune cognition, and culture 102
      8.3 Punctuated interpenetration: adapting to pathogenic hierarchy 105
      8.4 Implications for intervention 107

9 Cancer: a developmental perspective......................111
      9.1 Introduction 111
      9.2 Context 112
      9.3 Mutator dynamics 115
      9.4 Implications and speculations 117

10 Autoimmune disorders......................................121
      10.1 Introduction 121
      10.2 Nonorthogonal eigenmodes of immune cognition 124
      10.3 Circadian and other cycles..................................127
      10.4 The retina of the immune response 130
      10.5 Circadian-hormonal cycle synergism........................132
      10.6 The cognitive HPA axis 133
      10.7 Phase transitions of interacting information systems 135
      10.8 Autoimmune disease 136
      10.9 An application to asthma...................................138
      10.10 Images of pathogenic social hierarchy 139

11 Demoralization and obesity in Upper Manhattan 143
      11.1 Introduction 143
      11.2 Methods 144
      11.3 Data Analysis..147
      11.4 Modeling demoralization and chronic stress 148
      11.5 Results...149
      11.6 Discussion 154

	11.7 Conclusion 157

    11.7 Conclusion 157
    11.8 Chapter Appendix...158

12 Death at an early age: AIDS and related mortality in New York City 161
    12.1 Introduction 161
    12.2 Data and analysis 164
    12.3 Results...165
    12.4 Discussion 171
    12.5 Acknowledgment 178
    12.6 Chapter Appendix...178

13 Final thoughts 181

14 Mathematical appendix....................................183
    14.1 The Shannon Coding Theorem 183
    14.2 The 'tuning theorem' 185
    14.3 The Shannon-McMillan Theorem 187
    14.4 The Rate Distortion Theorem 190
    14.5 Groupoids 193
    14.6 Morse Theory 196
    14.7 Generalized Onsager Theory 198

15 References..201

Index 225

1
Introduction

1.1 Toward new tools

Researchers have begun exploring a de-facto cognitive paradigm for gene expression in which contextual factors determine the behavior of what Cohen calls a 'reactive system', not at all a deterministic, or even stochastic, mechanical process (e.g., Cohen, 2006; Cohen and Harel, 2007; Wallace and Wallace, 2008). The different approaches, while highly formal, are nonetheless much in the spirit of the pioneering efforts of Maturana and Varela (1980, 1992) who foresaw the essential role that cognitive process must play across a vast span of biological phenomena.

O'Nuallain (2008) has recently placed gene expression firmly in the realm of complex linguistic behavior, for which context imposes meaning, claiming that the analogy between gene expression and language production is useful both as a fruitful research paradigm and also, given the relative lack of success of natural language processing by computer, as a cautionary tale for molecular biology. First O'Nuallain argues that, at the orthographic or phonological level, depending on whether the language is written or spoken, we can map from phonetic elements to nucleotide sequence. His second claim is that Nature has designed highly ambiguous codes in both cases, and left disambiguation to the context.

He notes that, given our concern with the Human Genome Project (HGP) and its implications for human health, only 2% of diseases can be traced back to a straightforward genetic cause. As a consequence the HGP will have to be redone for a variety of metabolic contexts in order to establish a sound technology of genetic engineering (O'Nuallain and Strohman, 2007).

Here we investigate a broad class of probability models based on the asymptotic limit theorems of information theory that instantiate this perspective, finding a 'natural' means by which epigenetic context 'farms' gene expression in a sometimes punctuated manner via a kind of tunable catalysis. The models will be used to study how normal developmental modes can be driven by external context into pathological trajectories often expressed, in humans, as comorbid psychiatric and physical disorders. It appears possible to convert such models to powerful tools for data analysis, much as those based on the Central Limit Theorem can be converted to parametric statistics. A more formal version of the underlying mathematics can be found in Glazebrook and Wallace (2009).

We will begin with a summary of the biological contexts, then examine the popular spinglass model of development taken from neural network studies that we will ultimately generalize using a cognitive paradigm. The expanded approach permits import of tools and methods from statistical physics via the homology between information source uncertainty and free energy density, and this leads directly to the idea of epigenetic catalysis.

It is worth keeping in mind throughout the formal mathematics that Feynman's basic measure of information is simply the free energy needed to erase it (Feynman, 2000).

1.2 Epigenetic epidemiology

What we attempt is embedded in a large and lively intellectual context. Jablonka and Lamb (1995, 1998) have long argued that information can be transmitted from one generation to the next in ways other than through the base sequence of DNA. It can be transmitted through cultural and behavioral means in higher animals, and by epigenetic means in cell lineages. All of these transmission systems allow the inheritance of environmentally induced variation. Such Epigenetic Inheritance Systems are the memory structures that enable somatic cells of different phenotypes but identical genotypes to transmit their phenotypes to their descendants, even when the stimuli that originally induced these phenotypes are no longer present.

In chromatin-marking systems information is carried from one cell generation to the next because it rides with DNA as binding proteins or additional chemical groups that are attached to DNA and influence its activity. When DNA is replicated, so are the chromatin marks. One type of mark is the methylation pattern a gene carries. The same DNA sequence can have several different methylation patterns, each reflecting a different functional state. These alternative patterns can be stably inherited through many cell divisions.

Epigenetic inheritance systems are very different from the genetic system. Many variations are directed and predictable outcomes of environmental changes. Epigenetic variants are, in the view of Jablonka and Lamb, often, although not necessarily, adaptive. The frequency with which variants arise and their rate of reversion varies widely and epigenetic variations induced by environmental changes may be produced coordinatedly at several loci.

Parenthetically, some authors, e.g., Guerrero-Bosagna et al. (2005), disagree with the assumption of adaptiveness, inferring that input responsible for methylation effects simply produces a phenotypic variability then subject to selection. The matter remains open.

Jablonka and Lamb (1998) conclude that epigenetic systems may therefore produce rapid, reversible, co-ordinated, heritable changes. However such systems can also underlie non-induced changes, changes that are induced but non-adaptive, and changes that are very stable.

What is needed, they feel, is a concept of epigenetic heritability comparable to the classical concept of heritability, and a model similar to those used for measuring the effects of cultural inheritance on human behavior in populations.

Following a furious decade of research and debate, this perspective received much empirical confirmation. Backdahl et al. (2009), for example, write that epigenetic regulation of gene expression primarily works through modifying the secondary and tertiary structures of DNA (chromatin), making it more or less accessible to transcription. The sum and interaction of epigenetic modifications has been proposed to constitute an 'epigenetic code' that organizes the chromatin structure on different hierarchical levels (Turner, 2000). Modifications of histones include acetylation, methylation, phosphorylation, ubiquitination, and sumoylation, but other modifications have also been observed. Some are quite stable and play an important part in epigenetic memory, although DNA methylation is the only epigenetic modification that has maintenance machinery to preserve the marks through mitosis. This argues for DNA methylation to function as a form of epigenetic memory for the epigenome.

Codes and memory, of course, are inherent to any cognitive paradigm.

Jaenish and Bird (2003) argue that cells of a multicellular organism are genetically homogeneous but structurally and functionally heterogeneous owing to the differential expression of genes. Many of these differences in gene expression arise during development and are subsequently retained through mitosis. External influences on epigenetic processes are seen in the effects of diet on long-term diseases such as cancer. Thus, epigenetic mechanisms seem to allow an organism to respond to the environment through changes in gene expression. Epigenetic modifications of the genome provide a mechanism that allows the stable propagation of gene activity states from one generation of cells to the next. Because epigenetic states are reversible, they can be modified by environmental factors, and this may contribute to the development of abnormal responses. What needs to be explained, from their perspective, is the variety of stimuli that can bring about epigenetic changes, ranging from developmental progression and aging to viral infection and diet.

Jaenish and Bird conclude that the future will see intense study of the chains of signaling that are responsible for epigenetic programming. As a result, we will be able to understand, and perhaps manipulate, the ways in which the genome learns from experience.

Indeed, our central interest precisely regards the manner in which the asymptotic limit theorems of information theory constrain such chains of signaling, in the same sense that the Central Limit Theorem constrains sums of stochastic variates.

Crews et al. (2006, 2007) provide a broad overview of induced epigenetic change in phenotype, as do Guerrero-Bosagna et al. (2005), who focus particularly on early development. They propose that changes arising because of alterations in early development processes, in some cases environmentally induced, can appear whether or not such changes could become fixed and prosper in a population. They recognize two ways for this to occur, first by dramatically modifying DNA aspects in the germ line with transgenerational consequences – mutations or persistent epigenetic modifications of the genome – or by inducing ontogenetical variation in every generation, although not inheritance via the germ line. From their perspective, inductive environmental forces can act to create, through these means, new conformations of organisms which also implies new possibilities within the surrounding environment.

West-Eberhard (2003, 2005) takes a similar perspective that we will explore more fully in Chapter 4. She asserts that any new input, whether it comes from a genome, like a mutation or from the external environment, like a temperature change, a pathogen, or a parental opinion, has a developmental effect only if the preexisting phenotype is responsive to it. A new input causes a reorganization of the phenotype, what she calls a 'developmental recombination'. In that process phenotypic traits are expressed in new or distinctive combinations during ontogeny, or undergo correlated quantitative changes in dimensions. Developmental recombination can result in evolutionary divergence at all levels of organization.

According to West-Eberhard, individual development can be visualized as a series of branching pathways. Each branch point is a developmental decision, or switch point, governed by some regulatory apparatus, and each switch point defines a modular trait. Developmental recombination implies the origin or deletion of a branch and a new or lost modular trait. The novel regulatory response and the novel trait originate simultaneously, and their origins are inseparable events: there cannot be a change in the phenotype without an altered developmental pathway. West-Eberhard concludes that, contrary to the notion that mutational novelties have superior evolutionary potential, there are strong arguments for the greater evolutionary potential of environmentally induced novelties. An environmental factor can affect numerous individuals, whereas a mutation initially can affect only one.

In a similar way, Foley et al. (2009) argue that epimutation is estimated to be 100 times more frequent than genetic mutation and may occur randomly or in response to the environment. Periods of rapid cell division and epigenetic remodeling are likely to be most sensitive to stochastic or environmentally mediated epimutation. Disruption of epigenetic profile is a feature of most cancers and is speculated to play a role in the etiology of other complex diseases including asthma, allergy, obesity, type 2 diabetes, coronary heart disease, autism spectrum disorders, bipolar disorders, and schizophrenia.

They find evidence that a small change in the level of DNA methylation, especially in the lower range in an animal model, can dramatically alter expression for some genes. The timing of nutritional insufficiency or other environmental exposures may also be critical. In particular low-level maternal care was associated with developmental dysfunction and altered stress response in the young. Foley et al. emphasize the potential implications of such findings, given how widely stress is implicated in disease onset and relapse.

They especially note that when epigenetic status or change in status over time is the outcome, then models for either threshold-based dichotomies or proportional data will be required. Threshold models, defined by a given level or pattern of methylation or a degree of change in methylation over time, will, in their view, benefit from relevant functional data to identify meaningful thresholds.

A special contribution of our approach is that just such threshold behavior leads 'naturally' to a language-like 'dual information source' constrained by the necessary conditions imposed by information theory's asymptotic limit theorems, allowing development of statistical models of complicated cognitive phenomena, including but not limited to cognitive gene expression.

A recent review by Weaver (2009) focuses specifically on the epigenetic effects of glucocorticoids – stress hormones. In mammals, Weaver argues, the closeness or degree of positive attachment in parent-infant bonding and parental investment during early life has long-term consequences on development of inter-individual differences in cognitive and emotional development in the offspring. The long-term effects of the early social experience, he continues, particularly of the mother-offspring interaction, have been widely investigated. The nature of that interaction influences gene expression and the development of behavioral responses in the offspring that remain stable from early development to the later stages of life. Although enhancing the offspring's ability to respond according to environmental clues early in life can have immediate adaptive value, the cost, Weaver says, is that these adaptations serve as predictors of ill health in later life. He concludes that maternal influences on the development of neuroendocrine systems that underlie hypothalamic-pituitary-adrenal (HPA) axis and behavioral responses to stress mediate the relation between early environment and health in the adult offspring. In particular, he argues, exposure of the mother to environmental adversity alters the nature of mother-offspring interaction, which, in turn, influences the development of defensive responses to threat and reproductive strategies in the progeny.

In an updated review of epigenetic epidemiology, Jablonka (2004) finds it clear that the health and general physiology of animals and people can be affected not only by the interplay of their own genes and conditions of life, but also by the inherited effects of the interplay of genes and environment in their ancestors. These ancestral influences on health, Jablonka says, depend neither on inheriting particular genes, nor on the persistence of the ancestral environment.

Significantly, Bossdorf et al. (2008) invoke 'contexts' much like Baars' model of consciousness (Wallace, 2005a), and infer a need to expand the concept of variation and evolution in natural populations, taking into account several likely interacting ecologically relevant inheritance systems. Potentially, this may result in a significant expansion, though by all means not a negation, of the Modern Evolutionary Synthesis as well as in more conceptual and empirical integration between ecology and evolution.

More formally, Scherrer and Jost (2007a, b) use information theory arguments to extend the definition of the gene to include the local epigenetic machinery, something they characterize as the 'genon'. Their central point is that coding information is not simply contained in the coded sequence, but is, in their terms, *provided by* the genon that accompanies it on the expression pathway and controls in which peptide it will end up. In their view the information that counts is not about the identity of a nucleotide or an amino acid derived from it, but about the relative frequency of the transcription and generation of a particular type of coding sequence that then contributes to the determination of the types and numbers of functional products derived from the DNA coding region under consideration.

From our perspective the formal tools for understanding such phenomena involve asymptotic limit theorems constraining the dynamics of information sources – active systems that generate or 'provide' information – and these are respectively the Rate Distortion Theorem and its zero error limit, the Shannon-McMillan Theorem, described in the Mathematical Appendix.

1.3 Ecosystem resilience

Ecosystem resilience theory permits novel exploration of developmental psychiatric and chronic physical disorders resulting from dysfunctions of cognitive gene expression. Structured psychosocial stress, and similar noxious exposures, can, according to epigenetic epidemiology, write distorted images of themselves onto child growth, and, if sufficiently powerful, adult development as well, initiating a punctuated life course trajectory to characteristic forms of comorbid mind/body dysfunction that may be epigenetically heritable. For an individual, within the linked network of broadly cognitive physiological and mental subsystems, this occurs in a manner almost exactly similar to resilience domain shifts affecting a stressed ecosystem, suggesting that reversal or palliation may often be exceedingly difficult. Thus resilience theory may contribute significant new perspectives to the understanding, remediation, and prevention, of these debilitating conditions.

Ecosystem theorists recognize several different kinds of resilience (e.g., Gunderson, 2000). The first, that they call *engineering resilience*, since it characterizes machines and man-machine interactions, involves the rate at which a disturbed system returns to a presumed single, stable, equilibrium condition, following perturbation. From that limited perspective, a resilient system is one that quickly returns to its single stable state.

Not many biological phenomena – including those of human physiology and psychology – are actually resilient in this simplistic sense.

Holling's (1973) singular contribution was to recognize that sudden, highly punctuated, transitions between different, at best quasi-stable, domains of relation among ecosystem variates were possible, i.e., that more than one 'stable' state was possible for real ecosystems (Gunderson, 2000). Thus ecosystem resilience differs greatly from the engineering perspective.

We briefly examine a number of broadly cognitive mental and physiological phenomena, focusing on a spectrum of comorbid human developmental disorders that may, in fact, constitute complex, pathological, resilience domains. The usually irreversible and punctuated nature of path dependence in ecosystem domain shifts sheds new light on the therapeutic intractability of many such conditions.

It is important to note that punctuated biological phenomena are found across a great range of temporal scales (e.g., Eldredge, 1985; Gould, 2002). For example species appear 'suddenly' on a geologic time scale, persist relatively unchanged for a fairly long time, and then disappear suddenly, again on a geologic time scale. Evolutionary process is vastly speeded up in tumorigenesis, which nonetheless seems amenable to a roughly similar analysis (e.g., Wallace et al., 2003).

We invoke a model that adapts the asymptotic limit theorems of information theory to ecosystem resilience theory much in the way the central limit theorem is used in parametric statistical inference. The strength of such an approach is that it is almost independent of the detailed structure of the interacting information sources inevitably associated with both cognitive and ecosystem processes, important as such structure may be in other contexts.

We begin with a brief overview of some developmental pathologies and their comorbidities at both community and individual scales.

Comorbid disorders

Certain mental disorders, for example depression and substance abuse, and many physical conditions like lupus, coronary heart disease, hypertension, breast and prostate cancers, diabetes, obesity, and asthma, show marked regularities at the community level of organization according to the social constructs of race, ethnicity, socioeconomic status, and, appropriately, gender. Population-level structure in disease permits profound insight into etiology because, to the extent these are environmental disorders, the principal environment of humans is, in fact, other humans, moderated by a uniquely characteristic embedding cultural context (e.g., Durham, 1991; Richerson and Boyd, 2004). Thus culturally-sculpted social exposures are likely to be important at the individual, and critical at the population, levels of organization in the expression of certain mental disorders and a plethora of chronic diseases.

Further, mental disorders are often comorbidly expressed, both among themselves and with certain kinds of chronic physical disorder: Picture the obese, diabetic, depressed, anxious patient suffering from high blood pressure, asthma, coronary heart disease, and so on. Such comorbidity is the rule rather than the exception for the seriously ill.

As Cohen (2000) describes for autoimmune disease, however, the appearance of co- and antico- morbid conditions is, given the possibilities, rather surprisingly constrained to a relatively few often-recurring patterns.

Here we are interested in how cognitive submodules of human physiology and mental process may become synergistically linked with embedding, culturally structured, psychosocial stress to produce comorbid patterns of illness associated with mental disorder and chronic disease. A broad body of research suggests that many such disorders either have their roots in utero, as a stressed mother communicates environmental signals across the placenta, and programs her developing child's physiology, or else are initiated during early childhood (e.g., Bandelow et al., 2002; Barker, 2002; Barker et al., 2002, Coplan et al., 2005; Egle et al., 2002; Eriksson et al, 2000; Godfrey and Barker, 2001; Lalumiere et al., 2001; Mealey, 1995; Osmond and Barker, 2000; Repetti et al., 2000; G. Smith et al., 1998; T. Smith et al., 2002; Wright et al., 1998).

This pattern may affect underlying susceptibility to chronic infections or parasitic infestation as well as more systemic disorders (e.g., Wallace and Wallace, 2004).

The questions of central interest are the effects of stress on the interaction between mind and body over the life course. Stress is not often random in human societies, but is most frequently itself a socially constructed cultural artifact, very highly organized, having both a grammar and syntax. That is, certain stressors are meaningful in a particular developmental context, and others are not, with little or no long-term physiological effect. Stress is, then, often a kind of language.

We begin with a recitation of some cognitive submodules of human biology, in a large sense, that interact both with each other and with structured psychosocial stress. Next follows a brief exploration of cognition as a kind of language. Ultimately there emerges a generalized model based on Cohen's vision of autoimmune disease that accounts for a punctuated life trajectory of chronic comorbid psychiatric/physical disorder as involving a usually transient excited state of the system that becomes a pathologically permanent, or semi-permanent, zero-mode. Our particular innovation is to express this mechanism in terms of ecosystem resilience theory.

Some cognitive modules of human biology

In addition to cognitive gene expression, the central focus of our work, many other physiological systems have cognitive characteristics.

Immune function Atlan and Cohen (1998) have proposed an information-theoretic cognitive model of immune function and process, a paradigm incorporating cognitive pattern recognition-and-response behaviors analogous to those of the central nervous system. This work follows in a very long tradition of speculation on the cognitive properties of the immune system (e.g., Tauber, 1998; Podolsky and Tauber, 1998; Grossman, 1989, 1992, 1993a, b, 2000). We focus particularly on the Atlan/Cohen work, not because it conceived the idea of a cognitive immune system, but rather because it reinterprets cognitive process in terms of information in a particular manner.

From the Atlan/Cohen perspective, the meaning of an antigen can be reduced to the type of response the antigen generates. That is, the meaning of an antigen is functionally defined by the response of the immune system. The meaning of an antigen to the system is discernible in the type of immune response produced, not merely whether or not the antigen is perceived by the receptor repertoire. Because the meaning is defined by the type of response, there is indeed a response repertoire and not only a receptor repertoire.

To account for immune interpretation Cohen (1992, 2000) has reformulated the cognitive paradigm for the immune system. The immune system can respond to a given antigen in various ways, it has options. Thus the particular response we observe is the outcome of internal processes of weighing and integrating information about the antigen.

In contrast to Burnet's view of the immune response as a simple reflex, it is seen to exercise cognition by the interpolation of a level of information processing between the antigen stimulus and the immune response. A cognitive immune system organizes the information borne by the antigen stimulus within a given context and creates a format suitable for internal processing; the antigen and its context are transcribed internally into the 'chemical language' of the immune system.

The cognitive paradigm suggests a language metaphor to describe immune communication by a string of chemical signals. This metaphor is apt because the human and immune languages can be seen to manifest several similarities such as syntax and abstraction. Syntax, for example, enhances both linguistic and immune meaning.

Although individual words and even letters can have their own meanings, an unconnected subject or an unconnected predicate will tend to mean less than does the sentence generated by their connection.

The immune system creates a 'language' by linking two ontogenetically different classes of molecules in a syntactical fashion. One class of molecules are the T and B cell receptors for antigens. These molecules are not inherited, but are somatically generated in each individual. The other class of molecules responsible for internal information processing is encoded in the individual's germline.

Meaning, the chosen type of immune response, is the outcome of the concrete connection between the antigen subject and the germline predicate signals.

The transcription of the antigens into processed peptides embedded in a context of germline ancillary signals constitutes the functional 'language' of the immune system. Despite the logic of clonal selection, the immune system does not respond to antigens as they are, but to abstractions of antigens-in-context, consistent with O'Nuallain's (2008) perspective.

Tumor control Another cognitive submodule appears to be a tumor control mechanism that may include 'immune surveillance', but clearly transcends it. Nunney (1999) has explored cancer occurrence as a function of animal size, suggesting that in larger animals, whose lifespan grows as about the 4/10 power of their cell count, prevention of cancer in rapidly proliferating tissues becomes more difficult in proportion to size. Cancer control requires the development of additional mechanisms and systems to address tumorigenesis as body size increases – a synergistic effect of cell number and organism longevity. Nunney (1999, p. 497) concludes that this pattern may represent a real barrier to the evolution of large, long-lived animals and predicts that those that do evolve have recruited additional controls over those of smaller animals to prevent cancer.

Forlenza and Baum (2000) explore the effects of stress on the full spectrum of tumor control, ranging from DNA damage and control, to apoptosis, immune surveillance, and mutation rate. Elsewhere (R. Wallace et al., 2003) we argue that this elaborate tumor control strategy, particularly in large animals, must be at least as cognitive as the immune system that is one of its components. That is, some comparison must be made with an internal picture of a 'healthy' cell, and a choice made as to response: none, attempt DNA repair, trigger programmed cell death, engage in full-blown immune attack. This is, from the Atlan/Cohen perspective, the essence of cognition.

The HPA axis The hypothalamic-pituitary-adrenal (HPA) axis, a part of the general flight-or-fight system including the sympathoadrenomedullary system (SAM), is clearly cognitive in the Atlan/Cohen sense. Upon recognition of a new perturbation in the surrounding environment, memory and brain or emotional cognition evaluate and choose from several possible responses: no action needed, flight, fight, helplessness (i.e., flight or fight needed, but not possible). Upon appropriate conditioning, the HPA system, in coordination with the SAM axis, is able to accelerate the decision process, much as the immune system has a more efficient response to second pathogenic challenge once the initial infection has become encoded in immune memory. Certainly hyperreactivity in the context of post-traumatic stress disorder (PTSD) is a well-known example. Chronic HPA axis activation is deeply implicated in visceral obesity leading to diabetes and heart disease, via the leptin/cortisol diurnal cycle (e.g., Bjorntorp, 2001; Wallace and Wallace, 2005).

Blood pressure regulation Rau and Elbert (2001) review much of the literature on blood pressure regulation, particularly the interaction between baroreceptor activation and central nervous function. We paraphrase something of their analysis. The essential point, of course, is that unregulated blood pressure would be quickly fatal in any animal with a circulatory system, a matter as physiologically fundamental as tumor control. Much work over the years has elucidated some of the mechanisms involved. Increase in arterial blood pressure stimulates the arterial baroreceptors that in turn elicit the baroreceptor reflex, causing a reduction in cardiac output and in peripheral resistance, returning pressure to its original level. The reflex, however, is not actually this simple. It may be inhibited through peripheral processes, for example under conditions of high metabolic demand. In addition, higher brain structures modulate this reflex arc, for instance when threat is detected, and fight or flight responses are being prepared. This suggests, then, that blood pressure control cannot be a simple reflex, but is a broad and actively cognitive modular system that compares a set of incoming signals with an internal reference configuration, and then chooses an appropriate physiological level of blood pressure from a large repertory of possible levels.

Emotion Thayer and Lane (2000) summarize the case for what can be described as a cognitive emotional process. Emotions, in their view, are an integrative index of individual adjustment to changing environmental demands, an organismal response to an environmental event that allows rapid mobilization of multiple subsystems. Emotions are the moment-to-moment output of a continuous sequence of behavior, organized around biologically important functions. These lawful sequences have been termed behavioral systems by Timberlake (1994).

Emotions are self-regulatory responses that allow the efficient coordination of the organism for goal-directed behavior. Specific emotions imply specific eliciting stimuli, specific action tendencies including selective attention to relevant stimuli, and specific reinforcers. When the system works properly, it allows for flexible adaptation of the organism to changing environmental demands. Thus an emotional response represents a selection of an appropriate response and the inhibition of other less appropriate responses from a more or less broad behavioral repertoire of possible responses. Such 'choice' leads directly to something closely analogous to the Atlan and Cohen language metaphor.

Consciousness Although a Cartesian dichotomy between rational thought and emotion may be increasingly suspect, nonetheless humans, like many other animals, do indeed conduct conscious individual rational cognitive decision-making as most of us would commonly understand it. Various forms of dementia involve characteristic patterns of degradation in that ability. Dahaene and Naccache (2001) describe the global neuronal workspace model of consciousness, and our own extension of that model is available elsewhere (Wallace 2005a, 2006; Wallace and Fullilove, 2008), as is its explicit application to a spectrum of mental disorders (Wallace, 2005b).

Sociocultural network Humans are particularly noted for a hypersociality that inevitably enmeshes us all within group processes of decision, that is, collective cognitive behavior within a social network, tinged by an embedding shared culture (Wallace and Fullilove, 2008). For humans, culture is truly fundamental. Durham (1991) argues that genes and culture are two distinct but interacting systems of inheritance within human populations. Information of both kinds has influence, actual or potential, over behaviors, which creates a real and unambiguous symmetry between genes and phenotypes on the one hand, and culture and phenotypes, on the other. Genes and culture are best represented as two parallel lines or tracks of hereditary influence on phenotypes.

Cognition as 'language': an introduction

Atlan and Cohen (1998) argue that the essence of cognition is comparison of a perceived external signal with an internal, learned picture of the world, and then, upon that comparison, the choice of one response from a much larger repertoire of possible responses. We make a very general model of this process, a matter treated more formally in the next chapters.

A pattern of (broadly) sensory input is mixed in a systematic algorithmic way with internal ongoing activity to create a sequential path, $x = a_0, a_1, ...$, of composite signals a_i according to some algorithm. This path is then fed into a highly nonlinear, but otherwise unspecified, 'decision oscillator' $h(x)$ that generates an output which is an element of one of two (presumably) disjoint sets B_0 and B_1.

Thus we permit a graded response, supposing that if $h(x)$ is in B_0 the pattern is not recognized, and if $h(x)$ is in B_1, the pattern is recognized and some action takes place.

We are interested in composite paths x which trigger pattern recognition-and-response. That is, given a fixed initial state, such that $h(a_0) \in B_0$, we examine all possible subsequent paths x beginning with a_0 and leading to the event $h(x) \in B_1$.

For each positive integer n let $N(n)$ be the number of paths of length n which begin with some particular a_0 having $h(a_0)$ in B_0 and lead to the condition $h(x)$ in B_1. We shall call such paths 'meaningful' and assume $N(n)$ to be considerably less than the number of all possible paths of length n – pattern recognition-and-response is comparatively rare, and occurs according to rules. We further assume that the finite limit

$$H = \lim_{n \to \infty} \frac{\log[N(n)]}{n}$$

both exists and is independent of the path x. We call such a cognitive process ergodic (Ash, 1990).

After some development it is possible to define a stationary, ergodic information source as 'dual' to the ergodic cognitive process, as done in the next chapter.

A critical problem then becomes the choice of a 'normal' zero-mode language among a very large set of possible languages representing the (hyper- or hypo-) excited states accessible to the system (Wallace and Fullilove, 2008).

In sum, meaningful paths – creating an inherent grammar and syntax – have been defined entirely in terms of system response, as Atlan and Cohen (1998) propose.

A multiplicity of resilience topologies

An essential homology between information theory and statistical mechanics lies in the similarity of the expression for information source uncertainty above with the infinite volume limit of the free energy density. If $Z(K)$ is the statistical mechanics partition function derived from the system's Hamiltonian, then the free energy density is determined by the relation (e.g., Landau and Lifshitz, 2007)

$$F = \lim_{V \to \infty} -\frac{1}{K}\frac{\log[Z(K,V)]}{V} \equiv \frac{\log[\hat{Z}(K,V)]}{V}.$$

F is the free energy density, V the system volume and $K = 1/T$, where T is the system temperature.

It can be shown in some detail (e.g., Wallace and Wallace, 1998, 1999; Wallace, 2000, 2005a; Wallace and Fullilove, 2008; Rojdestvensky and Cottam, 2000; Feynman, 2000) that this systematic mathematical homology permits importation of powerful tools from statistical mechanics into information theory. Imposition of invariance under renormalization on the information splitting criterion H, or application of Landau's spontaneous symmetry breaking formalism, implies the existence of phase transitions analogous to learning plateaus or punctuated evolutionary equilibria in the relations between cognitive mechanism and external perturbation.

The homology also allows us to analyze the dynamical properties of interacting cognitive systems away from regions of phase change. The essential idea is that it is possible to define an entropy-analog using information source uncertainty as a 'free energy', and then define, in first order, a kind of empirical Onsager relation driving the dynamics of the system. See the Mathematical Appendix for details.

This formalism permits identification of three levels of (possibly interacting) topologies (Wallace and Fullilove, 2008). The first involves the dual information sources defined by equivalence classes of states connected by the high probability 'meaningful' paths leading from some base state in the construction above. States linked by such a path to the same base point form an equivalence class defining a groupoid that can be driven by spontaneous symmetry breaking through global parameter changes (Wallace and Fullilove, 2008).

Next, 'languages' that are similar in a precise sense form a dynamical equivalence class whose time dynamics are defined by analogs to generalized Onsager relations. These constitute formal dynamical manifolds and are associated with yet another groupoid.

Finally, paths within individual dynamical manifolds, individual dynamical paths can form directed homotopy equivalence classes.

This latter behavior mirrors the change in resilience mode when, for example, the normal forest state of low levels of spruce budworms shifts to a large-scale outbreak (Fleming and Shoemaker, 1992).

The effects of a crown forest fire on a spruce forest would be equivalent to a wholesale shifting of dynamics between different manifolds, in this formulation. Thus the set of possible dynamic manifolds, as a set of equivalence classes of behaviors, becomes itself a groupoid subject to linkages by crosstalk (Wallace and Fullilove, 2008; Glazebrook and Wallace, 2009).

Thus shifts in underlying topological equivalence classes, in this model, generalize the idea of resilience, and may involve both punctuated events within dynamical manifolds which are similar to conventional views of ecosystem resilience, and may involve in addition wholesale shifts between dynamical manifolds akin to clear-cutting or other catastrophe.

Our association of dual information sources with a spectrum of cognitive physiological phenomena thus implies that resilience analysis could be a new and very fundamental tool for understanding a broad range of physical and mental dysfunction in humans and their comorbid interactions.

In summary

The inference that a pattern of co- and antico- morbid mental and chronic physical disorder can represent the pathological permanent or semi-permanent establishment of an ordinarily atypical, or perhaps transient, state as a quasi-stable resilience mode is consistent, not only with recent ecosystem resilience theory, but with theorizing in both autoimmune disease and mental disorder.

Gilbert (2001), for example, uses an evolutionary approach to conclude that the relatively small number of evolved adaptive defense mechanisms, for example the flight-or-fight hypothalamic-pituitary-adrenal (HPA) axis, may become pathologically activated to produce mental disorder.

Jones and Blackshaw (2000) likewise argue that behavioral similarities between humans and animals show that many psychiatric states are distortions of evolved behavior, a perspective providing, in their view, a new etiological approach to psychiatry transcending current mainstream empirical and phenomenological approaches that are principally forms of symptom classification.

Although individual pathologies of both mind and body may predominate in particular cases, the analysis encompasses a broad swath of chronic diseases, emotional disorders, and classic cognitive dysfunction, in the context of the local sociocultural network so important to human biology (Richerson and Boyd, 2004).

Comorbidity may well be to medicine what the dirty open secret of punctuation in the fossil record has been to evolutionary theory (e.g., Gould, 2002), providing an opportunity for significant extension of our understanding and our ability to intervene against individual and population-level patterns of pathology.

For certain classes of mind/body symptomatology, early experiences of exposure to structured psychosocial stress – especially childhood poverty – appear particularly able to trigger identification of a highly atypical mode as the normal zero-reference state. This may occur in a punctuated manner, via a kind of epigenetic catalysis that we will describe in more detail below. Such a transition will often initiate a life course of co- or antico- morbid psychiatric and physical disorders. The characteristic pattern would involve individual and population-level comorbidity among obesity, asthma, diabetes, hypertension, depression, anxiety, substance abuse, ruthless or violent behaviors, coronary heart disease, certain cancers, and asthma or lupus – what might well be characterized as oppression disorder at the individual level. Obesity and its correlates will become a central focus of the later chapters of this book.

We will suggest that, at the individual scale, chronic, comorbid, mind/body dysfunctions are very much parallel to, and can be viewed as examples of, the more general ecosystem processes studied by resilience theory. Further analysis according to that theory, in particular the study of linkages across scale, could offer much to the understanding, treatment, and prevention, of developmental disorders and their comorbidities.

1.4 A first survey of the obesity pandemic in the US

The prevalence of obesity, one of the primary engines of chronic disease at both individual and larger scales, has been increasing for about a quarter century in the United States, is currently epidemic, and has become pandemic worldwide (e.g., Roth et al., 2004; Kimm and Obarzanek, 2002; Egger and Swinburn, 1997). Obesity poses a well-documented risk for many diseases and conditions, for example: cardiovascular disorders and type 2 diabetes. Recently, overweight/obesity has been associated with inflammation that may also contribute to the etiology of these diseases. During the past three decades, obesity spread geographically in the United States. While nationally obesity is concentrated in the poor states, especially the Southeast, in wealthy states like New York, obesity is concentrated in poor, primarily minority, urban neighborhoods. Together with adults, children in these poor neighborhoods have become fatter over time, as documented in subsequent waves of national household surveys. This increasing weight has put children at risk of obesity-related diseases at a relatively early age.

A vast literature has explored the relations between stress, adrenal hormones, leptin, and the hypothalamic-pituitary-adrenal axis. This literature suggests that individual stress, shaped by contextual stressors, plays a substantial role in establishing the behaviors and the endogenous characteristics that underlie obesity. The geography of overweight at the national and municipal levels then suggests that inter-group variability in contextual stressors may in part be responsible for the observed geographic differences in obesity. Studies have confirmed that contextual factors such as neighborhood conditions, instability of neighborhood social relations, and instability of residence impose stresses on households and individuals. Studies also have supported the hypothesis that neighborhood conditions, family function, and economic status have significant influences on fat deposition and the habits leading to it. We hypothesize that the obesity epidemic has roots in socioeconomic structure, area-level characteristics, and the relations between these variables. The surge in obesity in the past thirty years, the geographic patterns in distribution of obesity, and the relations between stress and obesity at the individual level all support this hypothesis, and are the particular subject of Chapter 7 below.

Numerous papers in the literature have documented the growth in overweight (body mass index>25) and obesity (BMI>30) in the United States since the 1970's. Between NHANES II (1976-1980) and NHANES III (1988-1994), a sharp increase in obesity occurred nationally for Americans of all ages, including in older pre-schoolers, 4-5 years old (Ogden et al., 1997). Over one-third of all adult Americans were obese in 1994. The most recent NHANES (1999-2000) confirmed the time trend: overweight prevalence increased from 56% to 64.5%, and age-adjusted obesity prevalence from 22.9 in the 1988-1994 period to 30.5 (Flegal et al., 2002). Nationally, overweight among preschool children increased from 5.8% in 1971-1974 to >10% in 1988-1994 (Ogden et al., 1997). The geography of overweight and obesity differs regionally with southeastern states showing the highest prevalences (Mokdad et al., 1999). Indeed, the changes in state-level prevalences over time suggest a diffusion of overweight and obesity from states of high prevalence to neighboring states (Mokdad et al., 2000) and a growing area of the country that has high prevalences of overweight and obesity (Mokdad et al., 1999). These patterns suggest a state-to-state diffusion of the risk behaviors (low levels of physical activity and high calorie consumption) that result in overweight/obesity. In New York City, the increase in overweight and obesity has been concentrated in poor, minority neighborhoods, and is highest among Latino children and African-Americans (Melnik et al., 1998). In New York City, overweight based on the 85th percentile in BMI was 37.5% among second graders and 31.7% among fifth graders (Melnik et al., 1998).

The epidemiology of overweight and obesity is complicated by a number of 'third variables' that frequently cooccur with obesity. For example, several surveys have examined class differences in body mass index, fat folds, waist-to-hip ratio, and other measures of overweight/obesity and body fat content.

It has been shown that school children of lower socioeconomic status are fatter (Wolfe et al., 1994). Few epidemiological surveys of either overweight/obesity or the diseases associated with them have assessed the role of race/ethnicity in obesity. For example, certain populations of Native Americans are known to have very high prevalences of overweight and obesity, such as the Pima Nation (Lindsay et al., 2002) although the explanation for this observation remains unclear.

Fewer epidemiologic studies have teased apart the different contribution that ethnicity makes from that of area conditions. Most American cities are not well integrated (Massey and Denton, 1992). Neighborhoods within large cities inhabited by African-Americans and by Latinos also have economic, social, and political characteristics: high poverty rates, high rates of forced residential instability, and low education – all characteristics that can be linked to health behaviors, and potentially to overweight, obesity and attendant diseases. To distinguish between ethnicity and area socioeconomic conditions as determinants of differences in overweight/obesity between ethnic groups would advance the public health science of obesity greatly and provide a direction for public policies to counteract the epidemic (e.g., D. Wallace et al., 2003, summarized in Chapter 11).

Fat is stored in specialized cells, the adipocytes. One of the most important findings of the 1990's is the glandular nature of adipocytes; they secrete hormones that shape the physiological processes of the body, from influencing the biological clock (Antonijevic et al., 1998), to directing whether T1 or T2 helper cells dominate (Lord et al., 1998). The best-studied of these hormones is leptin.

Leptin is released by fat cells after a meal (Houseknecht et al., 1998) and signals to the alimentary biological clock that the person has eaten and needs to eat no more. Leptin also influences and is influenced by the sleep/wake biological clock, peaking during the night whereas the adrenal hormone cortisol peaks during the day (Houseknecht et al., 1998; Casanueva, Dieguez, 1999). Leptin and cortisol maintain a dynamic balance; people with central abdominal obesity secrete cortisol faster than normal weighted people but clear it faster (Lottenberg et al., 1998). Cortisol is one of the adrenal hormones secreted in response to stress, especially to threatening stresses, a sign of the triggering of the generalized stress reaction of 'Fight or Flight' (Newcomer et al., 1998). The complementary circadian cycles of leptin and cortisol balance the need for sleep and for alertness against threats to survival. Furthermore, cortisol is accepted as a general marker of the status of the hypothalamus/pituitary/adrenal (HPA) axis (Ahlberg et al., 2002).

Recent work suggests links between stress, its attendant neurophysiological changes, and overweight/obesity. For example, people suffering from chronic stress-induced sleep pattern disruption metabolize their food differently from non-sufferers. The calories are channeled more into fat storage, and the fat is particularly deposited in the central abdominal area (Spiegel, Leproult, Van Cauter, 1999). The second round of Whitehall Studies revealed that hierarchical stress (stress imposed by lack of control over one's circumstances) was associated with higher BMI, central abdominal fat deposition, and higher rates of coronary heart disease and associated mortality (Brunner et al., 1997). The lower the occupational grade of the civil servants within the Whitehall cohort, the higher the prevalence of overweight/obesity, central abdominal fat deposition, and CHD.

Researchers in several other nations replicated and extended the results of the Whitehall Studies. Middle-aged Swedish men with markers of HPA axis activation were found to have higher prevalence of overweight/obesity and central abdominal fat deposition (Rosmond and Bjorntorp, 1998). Occupational class was found to associate strongly with overweight/obesity and waist-to-hip ratio (a measure of central abdominal fat deposition) among adults in several countries, including the United States (Marmot et al., 1998; Nakamura et al., 1998). Laboratory animal studies showed that leptin receptors are part of the hypothalamus (Iqbal et al., 2000). This suggests that fat cells send signals to the HPA, especially to the hypothalamus which, among other functions, regulates both the independent but related sleep/wake and eating biological clocks.

The second round of Whitehall Studies also revealed a relationship between hierarchical stress, fat, and inflammation. Occupational grade was associated with both fat deposition and markers of inflammation such as C-reactive protein and tumor necrosis factor (Brunner et al., 2002). Many previous physiological studies in both humans and animals had explored the role of the adrenal hormones, the pituitary, other parts of the neuroendocrine system and the adipohormones such as leptin in the activation of the immune system toward inflammation (inflammatory interleukins and white blood cells (Santos-Alvarez, Goberna, Sanchez-Margalet, 1999). For example: high leptin levels influence the T1/T2 helper cell balance toward dominance by T1, and the adrenal hormones toward T2 (Lord et al., 1998).

Leptin is also a proinflammant (Bullo et al., 2003); the adipohormones also trigger higher secretion of inflammatory interleukins and other immunomodulators (Santos-Alvarez, Goberna, Sanchez-Margalet, 1999). The present view of obesity is that of an inflammatory disease because of the strong association of excess fat with heightened inflammatory markers and incidence of inflammatory diseases. Stress-derived obesity has been strongly associated with markers of inflammation such as C-reactive protein (Brunner et al., 2002). Thus, several inflammatory diseases such as coronary heart disease, asthma, diabetes, and certain cancers are thought to have their roots in stress-derived obesity (Brunner et al., 2002, for coronary heart disease for example).

Abdominal obesity and visceral fat accumulation are particularly associated with disease, and have become the focus of much research on stress and its relation to the 'fight-or-flight' responses of the HPA axis. We paraphrase Bjorntorp (2001), who extensively summarizes the role of the HPA axis in physiological responses to stress. When the input of noxious signals is prolonged, the HPA axis reactivity changes from normal and relatively transient attempts to maintain homeostasis or allostasis with temporary peaks of cortisol secretion first, to a state of sensitization, which reacts with exaggerated cortisol secretion after a given input. This occurs during the most active phase of the HPA axis – the early morning in humans.

When repeated too often and with sufficient strength of the input, the first sign of malfunction is a delayed down-winding, so that cortisol secretion stays elevated for a prolonged period of time. Subsequently, the normal diurnal pattern is disrupted, and morning values tend to be lower. This subsequently develops into a low, steady, rigid diurnal cortisol secretion with little reactivity, a 'burned out' HPA. In parallel, the controlling, central glucocorticoid receptors become less efficient, and down-regulated.

Further challenges are followed by atrophy of the entire system, often found after long-term, severe hypercortisolaemia as in Cushing's syndrome, melancholic depression, post traumatic stress disorder (PTSD), and the aftermath of war. Much research shows that lowered sex steroid and growth hormone secretions have the same consequence, because of the insufficient counteraction against cortisol effects, and the combination of these abnormalities powerfully directs a larger than normal fraction of total body fat to visceral deposits.

In sum, increased activity of the HPA axis triggers inhibition of both the pituitary gonadal and growth hormone axes. Stress then may synergistically cause accumulation of visceral fat, via elevated cortisol secretion and decrease of sex steroid and growth hormones. Bjorntorp (2001) concludes in particular that the deposit of central body fat, which is closely correlated with general measures of obesity, can serve as a reasonable approximation to the long-term endocrine abnormalities associated with stress and often-repeated or chronic activation of the HPA axis. That is, stress literally writes an image of itself onto the body as visceral fat accumulation, first having written an image of itself onto the HPA axis. The phenomenon can be interpreted as the transmission of a structured signal between communicating systems, in a large sense, from the embedding psychosocial structure to an individual's HPA, as mathematically modeled in Chapter 7.

The opening salvo of the American Great Reform, Griscom's *On the Sanitary Conditions of the Laboring Classes of New York*, noted carefully the patterns of tuberculosis, drunkenness, and violence in relation to neighborhood conditions: housing overcrowding, housing design, residential instability, lack of education, working hours, work instability and insecurity, placement of groghouses within the multiple dwellings, etc. (Griscom, 1844). In his investigation of the draft riots of 1863, Griscom called these neighborhoods 'hives of sickness' and saw a seamless connection between physical health, mental health, and the spiritual receptivity for social control (Rosner, 1995).

The exploration between neighborhood conditions, public health, and public order ebbed and flowed in popularity among health scientists over the decades. The tuberculosis epidemics of various economic depressions often provoked the Tuberculosis League (ancestor of the American Lung Association) into active social health research. Thus, the connection between housing overcrowding and TB incidence was re-revealed decades ago (Wichen, 1934), long after it had been first commented on in the US. During the TB epidemic of the 1980's-early 1990's, this connection between housing overcrowding in poor neighborhoods and TB incidence was re-explored with tools of molecular biology to reveal that the strains of TB in cases in pre-school children were identical to the strains in adult cases in the same household and that the probability of within-household transmission was much greater when the overcrowding was at or above 2 persons per room (Drucker et al., 1995).

Although overcrowding within households, population density within neighborhoods, and residential instability that mixes population rapidly have obvious implications for infectious diseases such as tuberculosis and measles, contagious behaviors must also be considered in this context. Risk behaviors are actions that threaten the health and wellbeing of the perpetrator or of the perpetrator's social network. Risk behaviors include use of addictive substances, unsafe promiscuity, violence, and eating more than one's physical activity can burn off. Many risk behaviors have been well documented as contagious, i.e., patterns of behavior that spread along social networks and geographically. An early example is heroin use. Social research in the 1950's and 1960's documented the spread of heroin along social networks (Hunt and Chambers, 1976) and led to the term 'heroin epidemics'. In the past two decades, concentration of hard drug use in poor urban neighborhoods revealed the epicenters of the epidemics and the neighborhood conditions associated with them: high unemployment rates, low educational attainment among adults, residential instability and housing crisis, and resulting weak social structure for socializing the young (Currie, 1993).

A large and growing body of work examines the influence of neighborhood context on health, including, but not limited to, risk behaviors. Sampson et al. (1997) similarly found that violent crime rates within neighborhoods were associated with particular characteristics that render a community 'inefficacious'. Efficacious community (those that control social and physical conditions) are stable; the residents share values, trust each other, and engage in collective action. We cannot fully review the full range of the work on neighborhood context and health here; but other topics that have been studied include asthma (e.g., Wright et al., 1998), homicide (e.g., Fullilove et al., 2003; Wilson and Daly, 1997), cardiovascular health behaviors among youth (e.g., Lee and Cubbin, 2002), smoking among youth (Wilcox, 2003), and so on (e.g., Shinn and Toohey, 2003). Whole volumes on the topic now emerge with some regularity from the academic publishing houses (e.g., Kawachi and Berkman, 2002). A recurring finding of such studies is that, even after 'adjustment' for socioeconomic status (SES) and the social construct of 'race', health outcomes and risk behaviors show different systematic and determining patterns at the geographic scale of the immediate social and geographic neighborhood. As Shinn and Toohey (2003) point out, this finding is all the more remarkable as a result of the inevitable confounding of both 'race' and SES with residential location.

As Wilcox (2003) has noted, community-level contextual effects can operate directly on both groups and individuals as main effects, but can also condition the role of individual-level factors on individual behaviors, thus acting as moderating effects. Direct effects might include housing quality, availability, crowding, and safety, availability of financial resources (e.g., through 'redlining'), or quality of essential service delivery. Indirect effects on risk behaviors might be mediated by connectedness of youth to the social structures of older adults in that low connectedness between age cohorts can lead to enhanced connectedness within age cohorts, following Granovetter's (1973) 'strength of weak ties' arguments. Youth groups that strongly bound within, but not across, age cohort, suffer a disconnection from the larger community leading to a plethora of risks associated with the failure of 'collective social efficacy' in the supervision of adolescents (e.g., Fullilove et al., 2003; Wilcox, 2003; Sampson, 1990).

In a larger context, ecosystem theorists (e.g., Holling, 1992) have long argued for the essential role of meso-scale community structures acting as 'keystones', entraining processes at both larger and smaller scales of space, time, and population. That work has long recognized that individual or family (micro scale) and population level (macro scale) phenomena can both be driven by meso-scale (functional neighborhood) structure and process. Neighborhood effects in human health, illness, and risk behavior appear, from such a perspective, to be an expected, if striking, example of a systematic regularity found across many different ecosystems.

There are multiple sources of stress faced by individuals throughout their lifetime. While many of these stresses arise from individual life experiences, multiple stresses also arise from living conditions and from neighborhood context. With regard to the latter, a large literature on stressors in poor inner city neighborhoods exists. The psychobiosocial effects of witnessing murders (Li et al., 1998), exposure to drug selling (Li et al., 1999), feeling threatened by violence (Shumow et al., 1998), and infestation of one's home with vermin (Zahner et al., 1985) are well documented. Actual deprivation such as food and housing insecurity has also been well documented. There are a number of mechanisms that can be postulated to explain the relations between neighborhood-level characteristics and health behavior. We discuss a few of the most prominent of these hypotheses here.

First, overeating behavior may occur for the relief of varied states of stress and stress-related eating may ultimately contribute to overweight and obesity (Rhodes and Jason 1990; Lindenberg et al. 1994). Since previous research has shown stressful life events to occur with greater frequency in highly impoverished urban neighborhoods (Fang et al., 1998), overeating, like drug use, may be a way to cope with these events, consistent with stress reduction hypotheses. Experiences of discrimination and other negative stimuli may also be elevated in socially deprived neighborhoods, which in turn may lead to increased poor coping stress behaviors. For example, Anderson et al. (1990) found heightened police surveillance and police harassment in neighborhoods with substantial social strain. Similarly, neighborhood characteristics may increase levels of psychological distress (Aneshensel and Sucoff, 1996). For example, individuals living in multi-problem neighborhoods may be more likely to perceive risks such as crime, violence, noises, decrepit infrastructure and graffiti, which may lead to poor mental health. Poor mental health among residents of disadvantaged neighborhoods may also contribute to poor eating habits. Women in particular seem vulnerable to stress-related eating (Neumark-Sztainer et al., 1999) and their obesity is more closely associated with intake of sweets than is that of men's which is associated with alcohol and fatty foods (Rosmond and Bjorntorp, 1999).

Although stress is well-known to trigger eating of 'comfort foods' (Wilkinson, 1996), the relationship between stress and obesity/diabetes may also be indirect. Maternal stress is associated with low-weight birth, especially low-weight from prematurity (Copper et al., 1996). This low weight beginning may lead to overweight by age 7 and elevated risk of non-insulin dependent diabetes (NIDD) (Crowther et al., 1998). Although one subset of low-weight birth children remains at less than average weight through childhood, another set gains much weight, known as 'catchup weight gain' and has elevated risk of death from coronary heart disease (Eriksson et al., 1999). Low-weight babies are at higher risk of asthma (Brooks et al., 2001), which is also associated with obesity (Schachter et al., 2001).

Indeed, factors in the uterine environment are thought to contribute toward the susceptibility toward asthma (Annesi-Maesano, Moreau, Strachan, 2001); the uterine environment is also thought to influence susceptibility toward metabolic syndrome, central abdominal obesity, and NIDD (Phillips et al., 1999; Nyirenda and Seckl, 1998). For both asthma and NIDD, the prenatal influential factors are thought to be connected with socioeconomic status (Roberts, 1997), ethnicity (Collins et al., 1998a), neighborhood conditions (Collins et al., 1998b), and other structural factors determining maternal stress.

Second, adverse neighborhood conditions (e.g., disadvantage, limited collective efficacy) may undermine individuals' psychological resources and make poor health behavior more likely. Examples of the psychological resources that may be undermined include self-mastery. Wilson (1996) argues that concentrated unemployment in predominately black, ghetto communities has eroded a sense of personal self-mastery among the residents of these neighborhoods.

Third, it is possible that neighborhood disadvantage decreases social resources available to individuals, resulting in more limited assistance in coping with daily stresses, and fewer resources to overcome poor health behaviors and dysfunctional eating patterns once initiated.

Fourth, poor health behaviors may be related to neighborhood characteristics such as social norms through mechanisms unique to different neighborhoods and population groups. To this point, cultural norms have been linked to multiple health behaviors in previous studies (Linsky et al., 1986; Kaplan et al. 2001). For a pattern of behavior to be classed as a 'risk behavior', the indulging individual must be aware that it has consequences for health and/or safety for himself/herself or others. This knowledge is often gained from other members of the individual's community (Hoffman-Goetz, Mills, 1997). Thus, a community into which knowledge of risk posed by a particular indulgence has not penetrated may manifest a high prevalence of the risk behavior in all innocence. Flow of information into and within the community, sharing of values between the community and larger society, and sense of ability to control behavior are reflected in how rapidly and thoroughly mainstream society's understanding of the risks posed by particular indulgences is adopted by any one community. Information received from untrusted strangers such as governmental agencies is often ignored if it is not endorsed by a trusted member of the community (Santiago, 1994).

Fifth, differential availability of healthy foods may be directly associated with different eating patterns and different patterns of overweight/obesity.

Because the literature thus far has focused heavily on individual behaviors, many questions on the overweight/obesity epidemic remain unanswered.

The fact that overweight/obesity became an epidemic around 1979/1980 and has continued its epidemic dynamic indicates that population processes underlie this dynamic, not merely individuals and their behaviors. State-to-state spread of high prevalence (Mokdad et al., 1999) cannot occur without population processes driving the behavioral contagion. Yet, the whole chain of mechanisms is unknown: from the generation of susceptibility for the relevant risk behaviors in large subpopulations to the seeding of the risk behavior in the susceptible subpopulations to the role of social networks and their leadership in either facilitating or combating the risk behavior.

Concentration of an epidemic in certain areas necessarily implies that these are not self-contained zones where the disease is contained, but epicenters from which the disease spreads (Gould, 1993). A disease, whether infectious or behaviorally-mediated, cannot spread without sufficient density of susceptible individuals to support a contagious process (Bailey, 1975).

When stressors are socioeconomically structured, density of susceptibles follows the socioeconomic gradient and allows hierarchical, spatial, and network diffusion of the risk behavior, very much like spread of an infectious disease. The gradient in obesity prevalence nationally (Mokdad et al., 1999) as well as the subpopulation differences in obesity-related conditions such as diabetes (Harris, 1998) and cardiovascular disease (Greenlund et al., 1998) indicate hierarchical and spatially contagious modes of spread are operating in the US. The processes within levels of organization and linking levels of organization remain undefined for this particular set of risk behaviors.

In spite of tremendous attention devoted to the problem, the obesity epidemic has shown no sign of either slowing or declining. Diseases for which obesity is a risk factor such as Type 2 diabetes, certain cancers, and coronary heart disease/atherosclerosis have shown either large increases since the end of the 1970's or a failure to continue the previous pattern of decline (example: Greenlund et al., 1998 on cardiovascular risk factors among US women). The burden of overweight/obesity and their attendant comorbidities is enormous. Overweight/obesity is a risk factor for several diseases that rank high as causes of mortality: heart disease, diabetes, breast cancer, prostate cancer, cancer of the colon, and hypertension (Example: Goren et al., 2003). Obese people have a lower life expectancy than normal-weight people (Fontaine et al, 2003). Obese people require much greater frequency and duration of medical interventions, as well as number and dosage of drugs prescribed and duration of drug therapy (Bungum et al., 2003). Because the epidemic of overweight/obesity is concentrated in poor ethnic neighborhoods, it makes a major contribution to the well-documented health inequalities between rich and poor, between white and black, and between people of differing educational attainment.

Wallace et al. (1996) and Wallace and Fullilove (1999) have modeled how neighborhood and community factors can determine patterns of individual risk behavior by constraining the richness of the 'behavioral language' that community members can use to communicate along their social networks. Behaviors, according to this work, not just written or spoken language, constitute a principal means by which individuals communicate with their embedding social network. Marginalized communities have, according to this analysis, internal structures largely reflecting social, economic, historical, and even geographic constraints imposed upon them by the larger embedding society.

Opportunities for acquisition or attainment of symbols-of-worth from the larger, embedding society are limited for individuals within marginalized communities: certain occupations may be closed by 'grandfather' clauses of historical precedence, by rigid systems of examinations designed to select out those who do not use language in precisely the manner characterizing the majority population, and so on. Such exclusion from the larger, majority-defined, social arena of symbols of individual status or worth may thus trigger necessary substitutions that can become 'risk behaviors' under properly pathological circumstances.

That is, we suggest that an elaborate and characteristic behavioral repertory will develop within a marginalized subgroup, largely structured by an oppression which denies group members more 'conventional' symbols of individual and group worth, enabling it to carry out many social and individual purposes, much as a spoken or written language. Application of tools from information theory and the Large Deviations Program of applied probability permits creation of a model relating large-scale social constraints to sudden onset of a simplified behavioral language having a 'grammar' and 'syntax' of risk behavior as a tool for communicating along a 'noisy' social network affected by externally-imposed constraints (Wallace et al., 1996).

Chapter 11 examines contextual determinants of obesity-related behavior at the neighborhood scale. The analysis is based on an African-American and Dominican cohort of Upper Manhattan women who used the pre-natal clinics of Harlem Hospital and Columbia-Presbyterian Medical Center. Each participant's address was geocoded into one of the 27 health areas within the study zone. As described below, we developed an index of chronic community stress (ICCS) for the health areas, based on several stressors found to be strongly associated with health area incidence of low-weight births. When we grouped the women into their ICCS quintiles and plotted quintile average demoralization scores and BMI's against weighted average quintile ICCS, both health outcomes showed strong nonlinear relationships with ICCS. The average demoralization score exhibited an inverted 'U' shape such that the peak average demoralization score coincided with the middle ICCS. This relationship fit a model of stochastic resonance signal transduction very well: $R^2 = 0.75, P < 0.00001$. Average BMI also peaked in the middle ICCS quintile but plateau'd so that the BMIs of the fourth and fifth quintiles were not significantly different from that of the third. The average BMIs of the worst three quintiles for chronic community stress exceeded the cutoff for overweight. There was an obvious relationship between neighborhood conditions, demoralization, and BMI.

Chapter 11 explores the impact of community conditions on depression and body mass index of the Upper Manhattan cohort. Wallace and Wallace (1998b) described the impact of socioeconomic conditions at the county level on geographic patterns of asthma and diabetes mortality rates across eight very different metropolitan regions. Obesity is a risk factor of asthma, especially for women; asthma is a risk factor for obesity. Obesity is a strong risk factor of diabetes

Previous work (Wallace, Wallace, and Andrews, 1997) demonstrated that such non-infectious conditions as low-weight birth and violent crime could show hierarchical diffusion across metropolitan regions (regionalization). These conditions, like infectious diseases such as tuberculosis and AIDS, are markers of contagious social disintegration and are transmitted as part of contagious social disintegration along the socioeconomic connections between counties and between the core city and its dependent suburban counties.

Metropolitan regional asthma and diabetes mortality showed epidemic increases after 1980: slow increases before about 1985 and then rapid acceleration into the 1990's. When asthma mortality and diabetes mortality were analyzed with the regionalization/poverty model developed to analyze tuberculosis, AIDS, low birthweight and violent crime, the two showed different patterns. For six of the eight metropolitan regions studied, asthma mortality rate became regionalized after 1985, the time of the acceleration in mortality rate. Before 1985, county poverty rate was the main determinant of metropolitan geographic pattern in asthma mortality. Furthermore, measures of the strength of the association between the commuter index (the index of connection between counties) and asthma mortality rate among the eight metropolitan regions (F-ratio and R^2) were themselves associated with economic conditions of the core cities: unemployment rate per square mile of core city and number of people in poverty without public assistance.

Diabetes mortality presented a different picture. It was associated mainly with county poverty rates throughout the whole study period in all metropolitan regions but one, St. Louis. After 1985, diabetes mortality became regionalized over the St. Louis metropolitan region. Of all eight core cities, St. Louis had suffered the most economic and population decline. It had lost so much in population that the core city was only about half as populous as the major county by 1990.

Chapter 7 indicates, in some detail, that economic process at the national level, particularly the decline in manufacturing jobs, profoundly influences the obesity epidemic and mortality rate from such co-morbidities as diabetes and hypertension in the US, and begins the task of applying cognitive gene expression theory to the problem. Chapter 8 reconsiders coronary heart disease, and Chapter 9 applies the theory to hormonal cancers. Chapter 10 examines the intersection of the HPA and immune systems in terms of autoimmune disorders, focusing particularly on Lupus and asthma. Chapters 11 and 12 examine chronic diseases at neighborhood scales. While the theory seems to apply well to individual and simple aggregate scales of analysis, interactions between physiological systems within individuals, and between cognitive gene expression processes between individuals in tightly linked communities may display emergent and correlated patterns not easily accounted for in our present formalism.

We begin the formal work with a detailed reconsideration of the current defacto standard systems biology neural network-analog model of development, and proceed to its generalization.

2
Models of development

2.1 The spinglass model

Ciliberti et al.(2007a, b), culminating a long series of papers, apply the spinglass model from statistical physics to organisimal development in an evolutionary context. We summarize their results and look at some of the less obvious topological implications – in particular the mapping of disjoint directed homotopy classes of phenotype paths into interaction matrix space. We then extend the approach by applying a cognitive paradigm for gene expression first developed in Wallace and Wallace (2008). Analogs to phase change arguments for physical systems generate punctuated equilibrium evolutionary transitions in a 'highly natural' manner, even for the spinglass treatment, and a hierarchical extension permits incorporation of epigenetic effects as a kind of tunable catalysis.

The spinglass model of development assumes that N transcriptional regulators are represented by their expression patterns

$$\mathbf{S}(t) = [S_1(t), ..., S_N(t)]$$

at some time t during a developmental or cell-biological process and in one cell or domain of an embryo. The transcriptional regulators influence each other's expression through cross-regulatory and autoregulatory interactions described by a matrix $w = (w_{ij})$. For nonzero elements, if $w_{ij} > 0$ the interaction is activating, if $w_{ij} < 0$ it is repressing. w represents, in this model, the regulatory genotype of the system, while the expression state $\mathbf{S}(t)$ is the phenotype. These regulatory interactions change the expression of the network $\mathbf{S}(t)$ as time progresses according to a difference equation

$$S_i(t + \Delta t) = \sigma[\sum_{j=1}^{N} w_{ij} S_j(t)],$$

(2.1)

where Δt is a constant and σ a sigmodial function whose value lies in the interval $(-1, 1)$. In the spinglass limit σ is the sign function, taking only the values ± 1.

The networks of interest in the spinglass model are those whose expression state begins from a prespecified initial state $\mathbf{S}(0)$ at time $t = 0$ and converge to a prespecified stable equilibrium state $\mathbf{S}_\infty$. Such a network is termed *viable*, for obvious reasons.

After an elaborate and very difficult simulation, Ciliberti et al. (2007a) find that viable networks comprise a tiny fraction of possible ones. These could be widely scattered in the space of all possible networks and occupy disconnected islands in this space. However, direct computation indicates precisely the opposite. The metagraph of viable networks has one 'giant' connected component that comprises most or all viable networks. Any two networks in this component can be reached from one another through gradual changes of one regulatory interaction at a time, changes that never leave the space of viable networks, for this calculation.

In general, within the giant component, randomly chosen pairs of networks with the same phenotype will have vastly different organization, in terms of the matrix (w_{ij}).

Define $0 \leq d \leq 1$ as the fraction of genes that differ in their expression state between $\mathbf{S}_0$ and $\mathbf{S}_\infty$.

A typical result is that for $N = 5$ genes, $6 \leq M \leq 7$ total regulatory interactions, and $d = 0.4$, full enumeration finds a total of only 37,338 viable networks out of 6.3×10^7 possible ones (Ciliberti et al., 2007a). Long random walks through the space of viable networks, however, visit all but a very small fraction of the nodes of the metagraph, and this missing fraction decreases as N increases. Large N require elaborate Monte Carlo sampling for simulation, a difficult and computationally intensive enterprise.

In w-space Ciliberti et al. define a metric characterizing the distance between two network topologies as

$$D(w, w') = \frac{1}{2M_+} \sum_{i,j} |sign(w_{ij}) - sign(w'_{ij})|,$$

where M_+ is the maximum number of regulatory interactions, and sign(x)=± 1 depends on the sign of x, and is 0 for $x = 0$.

Several observations emerge directly.

[1] This approach is formally similar to spinglass neural network models of learning by selection, e.g., as proposed by Toulouse et al. (1986) nearly a generation ago. Subsequent work by Baars (1988, 2005), summarized in Dehaene and Naccache (2001), suggests that such models are simply not sufficient to the task of understanding high level cognitive function, and these have been

largely supplanted by complicated 'global workspace' concepts whose mathematical characterization is highly nontrivial (Atmanspacher, 2006).

[2] What Ciliberti et al. observe, in another idiom, is that in phenotype space, in **S**-space, the set of all paths associated with viable networks forms an equivalence class, closely analogous to the directed homotopy equivalence classes in the sense of Goubault and Raussen (2002) and Gobault (2003). Directed homotopy differs from simple homotopy (e.g., Lee, 2000) in that one uses paths from one point to another rather than loops, and seeks continuous deformations between them. See Wallace and Wallace (2008) for another discussion, in a biological context. Thus there is, in this spinglass model, a mapping from **S**-space into (w_{ij}) space, characterized by the metric D, that associates a unique simply connected component with each dihomotopy-like equivalence class of paths connecting two particular phenotype points. Indeed, the w-space component might well be treated according to standard homotopy arguments, i.e., using loops.

[3] What one does with homotopically simply connected components is patch them together to build larger, and more interesting, topological structures, using the Seifert-Van Kampen Theorem (SVKT) (e.g., Lee, 2000, Ch. 10). If paths within S-space are not continuously transformable into one another (if there are 'holes'), then several distinct dihomotopy classes will exist, e.g., as in figures 1 and 2 of Wallace and Wallace (2008), explored further below in terms of developmental critical periods and their 'shadows'. The obvious conjecture is that, under such a circumstance, very complex topological objects may lurk in w-space, not just the simply connected component discovered by by Ciliberti et al. These may, according to the SVKT, intersect as well as exist as isolated and disconnected sets.

If there are dihomotopy 'holes' in S-space, consequently reflected in disconnected patches in w-space, then punctuated transitions of various sorts may well become an evolutionary norm, (as in Gould, 2002), even for the spinglass model.

[4] A large and increasing body of work surrounding coupled cell networks invokes groupoids, a natural generalization of symmetry groups. As Dias and Stewart (2004) remark, until recently the abstract theory of coupled cell systems has mainly focused on the effects of symmetry in the network and the consequent formation of spatial and spatiotemporal patterns. The formal setting for this theory centers upon the symmetry group of the network.

Dias and Stewart conclude that analysis of robust patterns of synchrony in general coupled cell systems – that is, dynamics in which sets of cells behave identically as a consequence of the network topology – leads to the fruitful notion of the 'symmetry groupoid' of a coupled cell network. A groupoid is a generalization of a group, in which products of elements are not always defined. The symmetry groupoid of a coupled cell network is a natural algebraic formalization of the 'local symmetries' that relate subsets of the network to each other. In particular 'admissible' vector fields – those specified by the network topology – are precisely those that are equivariant under the action of the symmetry groupoid.

The Appendix provides a summary of standard material on groupoids that will be of later use.

[5] Both approaches can apparently be coarse-grained into a symbolic dynamics associated with (simple) information sources having particular grammar and syntax. The method is straightforward (e.g., Beck and Schlogl, 1995; McCauley, 1994). One could, thus, probably translate the spinglass results of Ciliberti et al. into symbolic dynamics, using groupoid methods to study the underlying topological objects.

[6] The spinglass model of development is abstracted from longstanding (if ultimately unsuccessful) attempts at similar treatments of neural networks involved in high level cognition (e.g., Jaeger et al., 2004; Mjolsness et al., 1991; Reinitz and Sharp, 1995; Sharp and Reinitz, 1998). Thus and consequently Ciliberti et al. are invoking an implicit cognitive paradigm for gene expression. Cognitive process, as the philosopher Fred Dretske (1994) eloquently argues, is constrained by the necessary conditions imposed by the asymptotic limit theorems of information theory. A little work produces a very general cognitive gene expression metanetwork structure recognizably similar to that found in Ciliberti et al. The massively parallel computations are hidden, somewhat, in the required empirical fitting of regression model analogs based on the asymptotic limit theorems of information theory rather than on the central limit theorem.

[7] A salient characteristic of high level cognitive process is precisely its inherent punctuation (e.g., Baars, 1988, 2005; Wallace, 2005a), and this emerges directly using an information theory approach via the famous homology between information and free energy. 'Simple' neural network analogs will inevitably have more difficulty replicating such behavior, but as discussed, the mapping of disjoint dihomotopy equivalence classes from phenotype sequence space to disjoint sets in interaction matrix space provides a straightforward example for spinglass models.

The next sections use information theory methods to make the transition from crossectional w-space into that of serially correlated sequences of phenotypes, expanding on the results of Wallace and Wallace (2008).

2.2 Shifting perspective: cognition as an information source

As described above, Atlan and Cohen (1998), in the context of a study of the immune system, argue that the essence of cognition is the comparison of a perceived signal with an internal, learned or inherited picture of the world, and then choice of a single response from a large repertoire of possible responses.

2.2 Shifting perspective: cognition as an information source

Such choice inherently involves information and information transmission since it always generates a reduction in uncertainty, as explained in Ash (1990, p. 21).

More formally, a pattern of incoming input – like the $\mathbf{S}(t)$ of equation (1) – is mixed in a systematic algorithmic manner with a pattern of internal ongoing activity – like the (w_{ij}) according to equation (1) – to create a path of combined signals $x = (a_0, a_1, ..., a_n, ...)$ – analogous to the sequence of $\mathbf{S}(t + \Delta t)$ of equation (1), with, say, $n = t/\Delta t$. Each a_k thus represents some functional composition of internal and external signals.

This path is fed into a highly nonlinear decision oscillator, h, a 'sudden threshold machine', in a sense, that generates an output $h(x)$ that is an element of one of two disjoint sets B_0 and B_1 of possible system responses. Let us define the sets B_k as

$$B_0 \equiv \{b_0, ..., b_k\},$$

$$B_1 \equiv \{b_{k+1}, ..., b_m\}.$$

Assume a graded response, supposing that if

$$h(x) \in B_0,$$

the pattern is not recognized, and if

$$h(x) \in B_1,$$

the pattern has been recognized, and some action $b_j, k + 1 \leq j \leq m$ takes place.

The principal objects of formal interest are paths x triggering pattern recognition-and-response. That is, given a fixed initial state a_0, examine all possible subsequent paths x beginning with a_0 and leading to the event $h(x) \in B_1$. Thus $h(a_0, ..., a_j) \in B_0$ for all $0 < j < m$, but $h(a_0, ..., a_m) \in B_1$.

For each positive integer n, let $N(n)$ be the number of high probability grammatical and syntactical paths of length n which begin with some particular a_0 and lead to the condition $h(x) \in B_1$. Call such paths 'meaningful', assuming, not unreasonably, that $N(n)$ will be considerably less than the number of all possible paths of length n leading from a_0 to the condition $h(x) \in B_1$.

While the combining algorithm, the form of the nonlinear oscillator, and the details of grammar and syntax are all unspecified in this model, the critical assumption which permits inference of the necessary conditions constrained by the asymptotic limit theorems of information theory is that the finite limit

$$H \equiv \lim_{n \to \infty} \frac{\log[N(n)]}{n}$$

(2.2)

both exists and is independent of the path x.

Define such a pattern recognition-and-response cognitive process as *ergodic*. Not all cognitive processes are likely to be ergodic in this sense, implying that H, if it indeed exists at all, is path dependent, although extension to nearly ergodic processes seems possible (Wallace and Fullilove, 2008).

Invoking the spirit of the Shannon-McMillan Theorem, whose content is described in more detail in the Appendix, as choice involves an inherent reduction in uncertainty, it is then possible to define an adiabatically, piecewise stationary, ergodic (APSE) information source $\mathbf{X}$ associated with stochastic variates X_j having joint and conditional probabilities $P(a_0, ..., a_n)$ and $P(a_n|a_0, ..., a_{n-1})$ such that appropriate conditional and joint Shannon uncertainties satisfy the classic relations

$$H[\mathbf{X}] = \lim_{n \to \infty} \frac{\log[N(n)]}{n} =$$

$$\lim_{n \to \infty} H(X_n|X_0, ..., X_{n-1}) =$$

$$\lim_{n \to \infty} \frac{H(X_0, ..., X_n)}{n+1}.$$

(2.3)

See the Mathematical Appendix for a summary of basic information theory results.

This information source is defined as *dual* to the underlying ergodic cognitive process.

Adiabatic means that the source has been parametized according to some scheme, and that, over a certain range, along a particular piece, as the parameters vary, the source remains as close to stationary and ergodic as needed for information theory's central theorems to apply. *Stationary* means that the system's probabilities do not change in time, and *ergodic*, roughly, that the cross sectional means approximate long-time averages. Between pieces it is necessary to invoke various kinds of phase transition formalisms, as described more fully in Wallace (2005a) and Wallace and Fullilove (2008).

2.2 Shifting perspective: cognition as an information source

Using the developmental vernacular of Ciliberti et al., we now examine paths in phenotype space that begins at some $\mathbf{S}_0$ and converges $n = t/\Delta t \to \infty$ to some other $\mathbf{S}_\infty$. Suppose the system is conceived at $\mathbf{S}_0$, and h represents (for example) reproduction when an appropriate phenotype $\mathbf{S}_\infty$ is reached. Thus $h(x)$ can have two values, i.e., B_0 not able to reproduce, and B_1, mature enough to reproduce. Then $x = (\mathbf{S}_0, \mathbf{S}_{\Delta t}, ..., \mathbf{S}_{n\Delta t}, ...)$ until $h(x) = B_1$.

Structure is now subsumed *within the sequential grammar and syntax of the dual information source* rather than within the cross sectional internals of (w_{ij})-space, a simplifying shift in perspective.

This transformation carries heavy computational burdens, as well as providing deeper mathematical insight.

First, the fact that 'viable' networks comprise a tiny fraction of all those possible emerges trivially from the spinglass formulation simply because of the 'mechanical' limit that the number of paths from $\mathbf{S}_0$ to a fixed $\mathbf{S}_\infty$ will always be far smaller than the total number of possible paths, most of which simply do not end on the target configuration.

A similar result also comes easily using Kolmogorov complexity (KC). Following Grunwald and Vitanyi (2003), the central idea of KC is to *examine the properties of the message sent* rather than of the information source sending it. The KC measure of a message is the minimal length of a program fed to a Universal Turing Machine (UTM) that replicates it. As Bennett (1982) puts it,

> A string is called 'algorithmically random' if it is not expressible as the output of a program much shorter than the string itself. A simple counting argument shows that, for any length N, most N-bit strings are algorithmically random. [For example], there are only enough $N - 10$ bit programs to describe at most $1/1024$ of all the N-bit strings.

From the information source perspective, that inherently subsumes a far larger set of dynamical processes than possible in a spinglass model or the set of things replicable by some theoretical (but unconstructable) Turing machine, the result is what Khinchin (1957) characterizes as the 'E-property' of a stationary, ergodic information source. This property is that, in the limit of infinitely long output, the classification of output strings into two sets is possible:

[1] a very large collection of gibberish which does not conform to underlying (sequential) rules of grammar and syntax, in a large sense, and which has near-zero probability, and

[2] a relatively small 'meaningful' set, in conformity with underlying structural rules, having very high probability.

The essential content of the Shannon-McMillan Theorem is that, if $N(n)$ is the number of meaningful strings of length n, then the uncertainty of an information source X can be defined as $H[X] = \lim_{n \to \infty} \log[N(n)]/n$, that can be expressed in terms of joint and conditional probabilities as in equation

(2.3) above. Proving these results for general stationary, ergodic information sources requires considerable sophisticated mathematical machinery, and does not at all emerge in the trivial manner of the spinglass or KC analyses (Cover and Thomas, 1991; Dembo and Zeitouni, 1998; Khinchin, 1957).

Thus the information source technique carries with it an inherent mathematical burden.

Second, information source uncertainty has an important heuristic interpretation that Ash (1990) describes as follows:

> ...[W]e may regard a portion of text in a particular language as being produced by an information source. The probabilities $P[X_n = a_n | X_0 = a_0, ... X_{n-1} = a_{n-1}]$ may be estimated from the available data about the language; in this way we can estimate the uncertainty associated with the language. A large uncertainty means, by the [Shannon-McMillan Theorem], a large number of 'meaningful' sequences. Thus given two languages with uncertainties H_1 and H_2 respectively, if $H_1 > H_2$, then in the absence of noise it is easier to communicate in the first language; more can be said in the same amount of time. On the other hand, it will be easier to reconstruct a scrambled portion of text in the second language, since fewer of the possible sequences of length n are meaningful.

This will prove important below.

Third, information source uncertainty is homologous with free energy density in a physical system, a matter having implications across a broad class of dynamical behaviors.

The free energy density of a physical system having volume V and partition function $Z(K)$ derived from the system's Hamiltonian – the energy function – at inverse temperature K is (e.g., Landau and Lifshitz, 2007)

$$F[K] = \lim_{V \to \infty} -\frac{1}{K} \frac{\log[Z(K,V)]}{V} = \lim_{V \to \infty} \frac{\log[\hat{Z}(K,V)]}{V},$$

(2.4)

where $\hat{Z} = Z^{-1/K}$.

The partition function for a physical system is the normalizing sum in an equation having the form

2.2 Shifting perspective: cognition as an information source

$$P[E_i] = \frac{\exp[-E_i/kT]}{\sum_j \exp[-E_j/kT]}$$

where E_i is the energy of state i, k a constant, and T the system temperature, and $P[E_i]$ is the probability of state i..

Feynman (2000), following the classic arguments of Bennett (1982, 1988) that present idealized machines using information to do work, concludes *the information contained in a message is most simply measured by the free energy needed to erase it*. Bennett's arguments are clever indeed, and Feynman's treatment of them is well worth reading.

Thus, according to this argument, source uncertainty is homologous to free energy density as defined above, i.e., from the similarity with the relation $H = \lim_{n \to \infty} \log[N(n)]/n$.

Ash's comment above then has an important corollary: If, for a biological system, $H_1 > H_2$, then, other things being equal, source 1 will require more metabolic free energy than source 2.

3
Groupoid symmetries

A formal equivalence class algebra, in the sense of the groupoid section of the Appendix, can now be constructed by choosing different origin and end points $\mathbf{S}_0, \mathbf{S}_\infty$ and defining equivalence of two states by the existence of a high probability meaningful path connecting them with the same origin and end. Disjoint partition by equivalence class, analogous to orbit equivalence classes for dynamical systems, defines the vertices of the proposed network of cognitive dual languages, much enlarged beyond the spinglass example. We thus envision a *network of metanetworks*, in the sense of Ciliberti et al. Each vertex then represents a different equivalence class of information sources dual to a cognitive process. This is an abstract set of metanetwork 'languages' dual to the cognitive processes of gene expression and development.

This structure generates a groupoid, in the sense of Weinstein (1996), as described in the Mathematical Appendix. States a_j, a_k in a set A are related by the groupoid morphism if and only if there exists a high probability grammatical path connecting them to the same base and end points, and tuning across the various possible ways in which that can happen – the different cognitive languages – parametizes the set of equivalence relations and creates the (very large) groupoid.

There is an implicit hierarchy. First, there is structure *within the system having the same base and end points*, as in Ciliberti et al. Second, there is a complicated groupoid structure defined by sets of dual information sources surrounding the variation of base and end points. We do not need to know what that structure is in any detail, but can show that its existence has profound implications.

We begin with the simple case, the set of dual information sources associated with a fixed pair of beginning and end states.

3.1 The first level

The spinglass model of Ciliberti et al. produced a simply connected, but otherwise undifferentiated, metanetwork of gene expression dynamics that could be traversed continuously by single-gene transitions in w-space. Taking the serial grammar/syntax model above, we find that not all high probability meaningful paths from $\mathbf{S}_0$ to $\mathbf{S}_\infty$ are actually the same. They are structured by the uncertainty of the associated dual information source, and that has a homological relation with free energy density.

Let us index possible dual information sources connecting base and end points by some set $A = \cup \alpha$. Argument by abduction from statistical physics is direct: Given metabolic energy density available at a rate M, and an allowed development time τ, let $K = 1/\kappa M\tau$ for some appropriate scaling constant κ, so that $M\tau$ is total developmental free energy. Then the probability of a particular H_α will be determined by the standard expression (e.g., Landau and Lifshitz, 2007),

$$P[H_\beta] = \frac{\exp[-H_\beta K]}{\sum_\alpha \exp[-H_\alpha K]},$$

(3.1)

where the sum may, in fact, be a complicated abstract integral.

This is just a version of the fundamental probability relation from statistical mechanics, as above. The sum in the denominator, the partition function in statistical physics, is a crucial normalizing factor that allows the definition of $P[H_\beta]$ as a probability.

A basic requirement, then, is that the sum/integral always converges. K is the inverse product of a scaling factor, a metabolic energy density rate term, and a characteristic development time τ. The developmental energy might be raised to some power, e.g., $K = 1/(\kappa(M\tau)^b)$, suggesting the possibility of allometric scaling.

Thus, in this formulation, there must be structure *within* a (cross sectional) connected component in the w-space of Ciliberti et al., determined in no small measure by available energy. Some dual information sources will be 'richer'/smarter than others, but, conversely, must use more metabolic energy for their completion.

3.2 The second level

The next generalization is crucial:

3.2 The second level

While we might simply impose an equivalence class structure based on equal levels of energy/source uncertainty, producing a groupoid in the sense of the Appendix (and possibly allowing a Morse Theory approach in the sense of Matsumoto, 2002, and Pettini, 2007), we can do more *by now allowing both source and end points to vary*, as well as by imposing energy-level equivalence. This produces a far more highly structured groupoid that we now investigate.

Equivalence classes define groupoids, by standard mechanisms (Brown, 1987; Golubitsky and Stewart, 2006; Weinstein, 1996). The basic equivalence classes – here involving both information source uncertainty level and the variation of S_0 and S_∞, will define transitive groupoids, and higher order systems can be constructed by the union of transitive groupoids, having larger alphabets that allow more complicated statements in the sense of Ash above.

Again, given an appropriately scaled, dimensionless, fixed, inverse available metabolic energy density rate and development time, so that $K = 1/\kappa M\tau$, we propose that the metabolic-energy-constrained probability of an information source representing equivalence class D_i, H_{D_i}, will again be given by

$$P[H_{D_i}] = \frac{\exp[-H_{D_i}K]}{\sum_j \exp[-H_{D_j}K]},$$

(3.2)

where the sum/integral is over all possible elements of the largest available symmetry groupoid. By the arguments of Ash above, compound sources, formed by the union of underlying transitive groupoids, being more complex, generally having richer alphabets, as it were, will all have higher free-energy-density-equivalents than those of the base (transitive) groupoids.

Let

$$Z_D \equiv \sum_j \exp[-H_{D_j}K].$$

(3.3)

We now define the *Groupoid free energy* of the system, F_D, at inverse normalized metabolic energy K, as

$$F_D[K] \equiv -\frac{1}{K}\log[Z_D[K]],$$

(3.4)

again following the standard arguments from statistical physics.

The groupoid free energy construct permits introduction of another important idea from statistical physics.

3.3 Spontaneous symmetry breaking

We have expressed the probability of an information source in terms of its relation to a fixed, scaled, available (inverse) metabolic free energy, seen as a kind of equivalent (inverse) system temperature. This gives a statistical thermodynamic path leading to definition of a 'higher' free energy construct – $F_D[K]$ – to which we now apply Landau's fundamental heuristic phase transition argument (Landau and Lifshitz, 2007; Pettini, 2007; Skierski et al, 1989).

The essence of Landau's insight was that certain phase transitions were usually in the context of a significant symmetry change in the physical states of a system, with one phase being far more symmetric than the other. A symmetry is lost in the transition, a phenomenon called spontaneous symmetry breaking. The greatest possible set of symmetries in a physical system is that of the Hamiltonian describing its energy states. Usually states accessible at lower temperatures will lack the symmetries available at higher temperatures, so that the lower temperature phase is less symmetric: The randomization of higher temperatures – in this case limited by available metabolic free energy – ensures that higher symmetry/energy states – mixed transitive groupoid structures – will then be accessible to the system. Absent high metabolic free energy, however, only the simplest transitive groupoid structures can be manifest. A full treatment from this perspective appears to require invocation of groupoid representations, no small matter (e.g., Bos, 2007; Buneci, 2003).

Somewhat more rigorously, the biological renormalization schemes of the Appendix to Wallace and Wallace (2008) may now be imposed on $F_D[K]$ itself, leading to a spectrum of phase transitions in the overall system of developmental information sources.

Most deeply, however, an extended version of Pettini's Morse-Theory-based topological hypothesis (Pettini, 2007) can now be invoked, i.e., that changes in underlying groupoid structure are a necessary (but not sufficient) consequence of phase changes in $F_D[K]$. Necessity, but not sufficiency, is important, as it, in theory, allows mixed groupoid symmetries.

The essential insight is that the single simply connected giant component of Ciliberti et al. is unlikely to be the full story, and that more complete models may be plagued – or graced – by highly punctuated dynamics.

Several matters are worth noting. First, Landau's spontaneous symmetry breaking arguments are perhaps the simplest approach possible here. The formal mathematical development requires invoking holonomy groups and groupoids, as in Glazebrook and Wallace, (2009).

Second, one need not be restricted to terms of the form $\exp[-H_j K]$, as any $f(H_j, K)$ such that the sum over j converges will serve, although the resulting 'thermodynamic' relations between variates of central interest may then be less elegant.

Third, there may be some allometric scaling tradeoff between metabolic energy rate and development time determined by a relation of the form $K \propto (\tau M)^\alpha$.

3.4 A biological example

This general formalism has broad implications for many biological processes. Here we briefly reconsider recent ideas on the origin of biological homochirality. On Earth, limited metabolic free energy density may have served as a low temperature-analog to 'freeze' the prebiotic system in the lowest energy state, i.e., the set of simplest homochiral transitive groupoids representing reproductive chemistries. These engaged in Darwinian competition until a single configuration survived. Subsequent path-dependent evolutionary process locked-in this initial condition. Astrobiological outcomes, in the presence of higher initial metabolic free energy densities, could well be considerably richer, for example, of mixed chirality. One result would be a complicated distribution of biological chirality across a statistically large sample of extraterrestrial stereochemistry, in marked contrast with recent published analyses predicting a racemic average.

Amino acids and the backbone of DNA/RNA in living things on Earth are found in only one of the two possible mirror-image states available to them. Respectively, the L-forms of amino acids primarily serve as the building blocks of proteins, and D-sugars form the DNA/RNA backbone (e.g., Fitz, et al. 2007). Attempts to replicate early conditions on Earth – Miller/Urey experiments – always produce 'racemic' mixtures having equal amounts of both possible amino acid symmetry forms. This fundamental conundrum was recently addressed by Gleiser et al. (2008), in a computational intensive study adapting Sandars' (2003) 'toy model' of polymerization. They conclude that other planetary platforms in this solar system and elsewhere could have developed an opposite chiral bias to that of Earth. As a consequence, a statistically large sampling of extraterrestrial stereochemistry would, in their view, be necessarily racemic on average.

It is possible to make a more direct attack on the problem based on the homology between free energy density and information source uncertainty. We argue, via the statistical thermodynamic constructions above, that available metabolic energy could well have been the principal determining environmental influence, and that, as a consequence of groupoid symmetries associated with stereochemical structure, a statistically large sampling of extraterrestrial stereochemistries could well be far more complex than Gleiser et al. propose, i.e., not racemic on average.

The argument is straightforward and involves several basic ideas:

[1] Reproducing molecular codes, in the largest sense, themselves constitute information sources that are Darwinian individuals, subject to variation, selection, and chance extinction in the sense of Gould (2002).

[2] Enantiomeric forms of molecules constitute equivalence classes that can be represented by groupoid, rather than group, symmetries, leading to a groupoid version of Landau's classic phenomenological model for phase transition and its extension via Pettini's (2007) topological hypothesis. The necessity of using groupoid methods in stereochemistry has long been recognized, and will not be reviewed here (e.g., Dornberger-Schiff and Grell-Niemann, 1961; Klemperer 1973; Nourse 1975; Sadanaga and Ohsumi, 1979; Fichtner, 1986; Yamamoto and Ishihara, 1988; Cayron, 2006, 2007). For a tutorial on groupoid methods, again see the Mathematical Appendix.

[3] Groupoid symmetries and available metabolic free energy are, as a consequence of the Darwinian individuality of coding schemes, contexts for, rather than determinants of, the resulting evolutionary processes, including punctuated equilibrium. That is, they define the banks between which the evolutionary glacier flows – sometimes slowly, sometimes in a sudden avalanche.

We begin by classifying the available molecules in our prebiotic soup by their underlying stereochemistries, and allow the reproductive systems to, for purposes of initial classification, reflect those stereochemical equivalence classes. Interactions between stereochemical equivalence classes can be used to classify higher order structures.

Equivalence classes define groupoids, as outlined in the Mathematical Appendix. The basic equivalence classes will define transitive groupoids, and higher order systems can be constructed by the union of transitive groupoids, having larger chemical alphabets that allow more complicated statements in the sense of Ash above.

Given an appropriately scaled, dimensionless, fixed, inverse available metabolic energy density K, we propose that the metabolic-energy-constrained probability of a reproductive information source representing stereochemical equivalence class D_i, H_{D_i}, will be given by equation (3.2), where the sum is over all possible elements of the largest available symmetry groupoid. By the arguments above, compound sources, formed by the union of (interaction of species from) underlying transitive groupoids, being more complex, will all have higher free-energy-density-equivalents than those of the base (transitive) groupoids.

Now apply equations (3.3) and (3.4) and the spontaneous symmetry breaking argument.

Absent high metabolic free energy densities, only the simplest transitive groupoid structures can be manifest in short development times, i.e., those associated with the simplest stereochemistries.

The necessity, but not sufficiency, of Pettini's topological hypothesis eventually allows mixed symmetries, e.g., L-forms of amino acids working in concert with the D-sugar DNA/RNA backbone.

What, in one sense, fatally compromises this analysis, but in a more fundamental way completes it, is that the reproductive chemical strategies represented by the H_{D_j} of equation (3.2) are not merely passive actors. Quite the contrary, they are full-scale Darwinian individuals in the sense of Gould (2002), subject to variation, selection, and chance extirpation. Thus, given sufficient initial metabolic energy density, there is no inherent reason why higher order, non-transitive, groupoid reproductive chemical systems – of mixed chirality – might not prevail, particularly in view of the Ash quotation of Section 2.2. That is, one can 'say' more in a shorter time using a richer reproductive language, and this might well have selective value. Thus we may, if this model is correct, expect to observe some surprising astrobiological reproductive stereochemistries, in contrast to the simple 'racemic' conclusion of Gleiser et al. (2008).

The corollary to this argument is that initial metabolic free energy density on Earth may just not have been sufficient to activate non-homochiral reproductive chemistries, and that the two possible amino acid systems, L, D, engaged in a competition through which one prevailed. Subsequent path-dependent evolutionary lock-in produced the ultimate result.

Again, groupoid symmetries and available metabolic free energy are, as a consequence of the Darwinian individuality of reproductive coding schemes, contexts for, rather than determinants of, evolutionary process, including punctuated equilibrium. As said, they are the banks between which the prebiotic evolutionary glacier flowed – sometimes slowly, and sometimes in sudden advance.

4
Epigenetic catalysis

4.1 The basic idea

Incorporating the influence of embedding contexts – epigenetic effects – on development is most elegantly done by invoking the Joint Asymptotic Equipartition Theorem (JAEPT) (Cover and Thomas, 1991). For example, given an embedding contextual information source, say Z, that affects development, then the dual cognitive source uncertainty H_{D_i} is replaced by a joint uncertainty $H(X_{D_i}, Z)$. The objects of interest then become the jointly typical dual sequences $y^n = (x^n, z^n)$, where x is associated with cognitive gene expression and z with the embedding context. Restricting consideration of x and z to those sequences that are in fact jointly typical allows use of the information transmitted from Z to X as the splitting criterion.

One important inference is that, from the information theory 'chain rule' (Cover and Thomas, 1991),

$$H(X, Y) = H(X) + H(Y|X) \leq H(X) + H(Y),$$

while there are approximately $\exp[nH(X)]$ typical X sequences, and $\exp[nH(Z)]$ typical Z sequences, and hence $\exp[n(H(X) + H(Z))]$ independent joint sequences, there are only about $\exp[nH(X, Z)] \leq \exp[n(H(X) + H(Z))]$ jointly typical sequences, so that the effect of the embedding context, in this model, is to lower the *relative* free energy of a particular developmental course.

Thus the effect of epigenetic regulation is to channel development into pathways that might otherwise be inhibited by an energy barrier. Hence the epigenetic information source Z acts as a *tunable catalyst*, a kind of second order cognitive enzyme, to enable and direct developmental pathways. This result permits hierarchical models similar to those of higher order cognitive neural function that incorporate Baars' contexts in a natural way (Wallace and Wallace, 2008; Wallace and Fullilove, 2008).

It is worth emphasizing that this is indeed a relative energy argument, since, metabolically, two systems must now be supported, i.e., that of the 'reaction' itself and that of its catalytic regulator. 'Programming' and stabilizing inevitably intertwined, as it were.

This elaboration allows a spectrum of possible 'final' phenotypes, what Gilbert (2001) calls developmental or phenotype plasticity. Thus gene expression is seen as, in part, responding to environmental or other, internal, developmental signals.

As described previously, West-Eberhard (2005) argues that any new input, whether it comes from the genome, like a mutation, or from the external environment, like a temperature change, a pathogen, or a parental opinion, has a developmental effect only if the preexisting phenotype is responsive to it. A new input causes a reorganization of the phenotype, or 'developmental recombination.' In developmental recombination, phenotypic traits are expressed in new or distinctive combinations during ontogeny, or undergo correlated quantitative change in dimensions. Developmental recombination can result in evolutionary divergence at all levels of organization.

Individual development can be visualized as a series of branching pathways. Each branch point, according to West-Eberhard, is a developmental decision, or switch point, governed by some regulatory apparatus, and each switch point defines a modular trait. Developmental recombination implies the origin or deletion of a branch and a new or lost modular trait. It is important to realize that the novel regulatory response and the novel trait originate simultaneously. Their origins are, in fact, inseparable events. There cannot, West-Eberhard concludes, be a change in the phenotype, a novel phenotypic state, without an altered developmental pathway.

These mechanisms are accomplished in our formulation by allowing the set B_1 in section 2.2 to span a distribution of possible 'final' states $\mathbf{S}_\infty$. Then the groupoid arguments merely expand to permit traverse of both initial states and possible final sets, recognizing that there can now be a possible overlap in the latter, and the epigenetic effects are realized through the joint uncertainties $H(X_{D_i}, Z)$, so that the epigenetic information source Z serves to direct as well the possible final states of X_{D_i}.

Again, Scherer and Jost (2007a, b) use information theory arguments to suggest something similar to epigenetic catalysis, finding the information in a sequence is not contained in the sequence but has been provided by the machinery that accompanies it on the expression pathway. That work does not, however, invoke a cognitive paradigm, its attendant groupoid symmetries, or the homology between information source uncertainty and free energy density that drives dynamics.

The mechanics of channeling can be made more precise as follows.

4.2 Rate Distortion dynamics

Real time problems, like the crosstalk between epigenetic and genetic structures, are inherently rate distortion problems, and the interaction between biological structures can be restated in communication theory terms. Suppose a sequence of signals is generated by a biological information source Y having output $y^n = y_1, y_2, ...$ This is 'digitized' in terms of the observed behavior of the system with which it communicates, say a sequence of observed behaviors $b^n = b_1, b_2, ...$ The b_i happen in real time. Assume each b^n is then deterministically retranslated back into a reproduction of the original biological signal,

$$b^n \to \hat{y}^n = \hat{y}_1, \hat{y}_2, ...$$

The information source Y is the epigenetic Z, and B is X_{D_i}, but the terminology used here is more standard (Cover and Thomas, 1991).

Define a distortion measure $d(y, \hat{y})$ to compare the original to the retranslated path. Many distortion measures are possible, as described in the Mathematical Appendix.

The distortion between *paths* y^n and $\hat{y}^n$ is defined as

$$d(y^n, \hat{y}^n) \equiv \frac{1}{n} \sum_{j=1}^{n} d(y_j, \hat{y}_j).$$

A remarkable fact of the Rate Distortion Theorem is that *the basic result is independent of the exact distortion measure chosen* (Cover and Thomas, 1991; Dembo and Zeitouni, 1998).

Suppose that with each path y^n and b^n-path retranslation into the y-language, denoted $\hat{y}^n$, there are associated individual, joint, and conditional probability distributions

$$p(y^n), p(\hat{y}^n), p(y^n, \hat{y}^n), p(y^n|\hat{y}^n).$$

The average distortion is defined as

$$D \equiv \sum_{y^n} p(y^n) d(y^n, \hat{y}^n).$$

(4.1)

It is possible, using the distributions given above, to define the information transmitted from the Y to the $\hat{Y}$ process using the Shannon source uncertainty of the strings:

$$I(Y,\hat{Y}) \equiv H(Y) - H(Y|\hat{Y}) = H(Y) + H(\hat{Y}) - H(Y,\hat{Y}),$$

(4.2)

where $H(...,...)$ is the joint and $H(...|...)$ the conditional uncertainty.

If there is no uncertainty in Y given the retranslation $\hat{Y}$, then no information is lost, and the systems are in perfect synchrony.

In general, of course, this will not be true.

The *rate distortion function* $R(D)$ for a source Y with a distortion measure $d(y,\hat{y})$ is defined as

$$R(D) = \min_{p(y,\hat{y}); \sum_{(y,\hat{y})} p(y)p(y|\hat{y})d(y,\hat{y}) \leq D} I(Y,\hat{Y}).$$

(4.3)

The minimization is over all conditional distributions $p(y|\hat{y})$ for which the joint distribution $p(y,\hat{y}) = p(y)p(y|\hat{y})$ satisfies the average distortion constraint (i.e., average distortion $\leq D$).

The *Rate Distortion Theorem* states that $R(D)$ is the minimum necessary rate of information transmission that ensures communication does not exceed average distortion D. Thus $R(D)$ defines a minimum necessary channel capacity. Cover and Thomas (1991) and Dembo and Zeitouni (1998) provide details. The rate distortion function has been explicitly calculated for a number of simple systems.

Recall, now, the relation between information source uncertainty and channel capacity:

$$H[\mathbf{X}] \leq C,$$

(4.4)

where H is the uncertainty of the source X and C the channel capacity, defined according to the relation

$$C \equiv \max_{P(X)} I(X|Y).$$

(4.5)

X is the message, Y the channel, and the probability distribution $P(X)$ is chosen so as to maximize the rate of information transmission along a Y.

Finally, recall the analogous definition of the rate distortion function above, again an extremum over a probability distribution.

Recall, again, equation 2.4 and the results of Chapter 3, i.e., that the free energy of a physical system at a normalized inverse temperature-analog $K = 1/\kappa T$ is defined as $F(K) = -\log[Z(K)]/K$ where $Z(K)$ the partition function defined by the system Hamiltonian. More precisely, if the possible energy states of the system are a set $E_i, i = 1, 2, \ldots$ then, at normalized inverse temperature K, the probability of a state E_i is determined by the relation $P[E_i] = \exp[-E_i K] / \sum_j \exp[-E_j K]$.

The partition function is simply the normalizing factor.

Applying this formalism, it is possible to extend the rate distortion model by describing a probability distribution for D across an ensemble of possible rate distortion functions in terms of available free metabolic energy, $K = 1/\kappa M \tau$.

The key is to take the $R(D)$ as representing energy as a function of the average distortion. Assume a fixed K, so that the probability density function of an average distortion D, given a fixed K, is then

$$P[D, K] = \frac{\exp[-R(D)K]}{\int_{D_{min}}^{D_{max}} \exp[-R(D)K]dD}.$$

(4.6)

Thus lowering K in this model rapidly raises the possibility of low distortion communication between linked systems.

We define the *rate distortion partition function* as just the normalizing factor in this equation:

$$Z_R[K] \equiv \int_{D_{min}}^{D_{max}} \exp[-R(D)K]dD,$$

(4.7)

again taking $K = 1/\kappa M\tau$.

We now define a new free energy-analog, the *rate distortion free-energy*, as

$$F_R[K] \equiv -\frac{1}{K}\log[Z_R[K]],$$

(4.8)

and apply Landau's spontaneous symmetry breaking argument to generate punctuated changes in the linkage between the genetic information source X_{D_i} and the embedding epigenetic information source Z. Recall that Landau's insight was that certain phase transitions were usually in the context of a significant symmetry change in the physical states of a system.

Again, the biological renormalization schemes of the Appendix to Wallace and Wallace (2008) may now be imposed on $F_R[K]$ itself, leading to a spectrum of highly punctuated transitions in the overall system of interacting biological substructures.

Since $1/K$ is proportional to the embedding metabolic free energy, we assert that

[1] the greatest possible set of symmetries will be realized for high developmental metabolic free energies, and

[2] phase transitions, related to total available developmental metabolic free energy, will be accompanied by fundamental changes in the final topology of the system of interest – phenotype changes – recognizing that evolutionary selection acts on phenotypes, not genotypes.

The relation $1/K = \kappa M\tau$ suggests the possibility of evolutionary tradeoffs between development time and the rate of available metabolic free energy.

4.3 More topology

It seems possible to extend this treatment using standard topological arguments.

Taking $T = 1/K$ in equations 3.2 and 4.6 *as a product of eigenvalues*, we can define it as the determinant of a Hessian matrix representing a Morse

Function, f, on some underlying, background, manifold, $\mathcal{M}$, characterized in terms of (as yet unspecified) variables $\mathcal{X} = (x^1, ..., x^n)$, so that

$$1/K = \det(\mathcal{H}_{i,j}),$$

$$\mathcal{H}_{i,j} \equiv \partial^2 f / \partial x^i \partial x^j.$$

(4.9)

Again, see the Appendix for a brief outline of Morse Theory.

Thus κ, M, and the development time τ are seen as eigenvalues of $\mathcal{H}$ on the manifold $\mathcal{M}$ in an abstract space defined by some set of variables $\mathcal{X}$.

By construction $\mathcal{H}$ has everywhere only nonzero, and indeed, positive, eigenvalues, whose product thereby defines T as a generalized volume. Thus, and accordingly, all critical points of f have index zero, that is, no eigenvalues of $\mathcal{H}$ are ever negative at any point, and hence at any critical point $\mathcal{X}_c$ where $df(\mathcal{X}_c) = 0$.

This defines a particularly simple topological structure for $\mathcal{M}$: If the interval $[a, b]$ contains a critical value of f with a single critical point $\mathcal{X}_c$, then the topology of the set $\mathcal{M}_b$ defined above differs from that of $\mathcal{M}_a$ in a manner determined by the index i of the critical point. $\mathcal{M}_b$ is then homeomorphic to the manifold obtained from attaching to $\mathcal{M}_a$ an i-handle, the direct product of an i-disk and an $(m-i)$-disk.

One obtains, in this case, since $i = 0$, the two halves of a sphere with critical points at the top and bottom (Matsumoto, 2002; Pettini, 2007). This is, as in Ciliberti et al. (2007a, b), a simply connected object. What one does then is to invoke the Seifert-Van Kampen Theorem (SVKT, Lee, 2000) and patch together the various simply connected subcomponents to construct the larger, complicated, topological object representing the full range of possibilities.

The physical natures of κ, M, and τ thus impose constraints on the possible complexity of this system, in the sense of the SVKT.

4.4 Inherited epigenetic memory

The cognitive paradigm for gene expression invoked here requires an internal picture of the world against which incoming signals are compared – algorithmically combined according to the rules of Section 2.2 – and then fed into a sharply stepwise decision oscillator that chooses one (or a few) action(s) from a much large repertoire of possibilities. Memory is inherent, and much recent work, as described in the introduction, suggests that epigenetic memory is indeed heritable.

The abduction of spinglass and other models from neural network studies to the analysis of development and its evolution carries with it the possibility of more than one system of memory. What Baars called 'contexts' channeling high level animal cognition may often be the influence of cultural inheritance, in a large sense. Our formalism suggests a class of statistical models that indeed greatly generalize those used for measuring the effects of cultural inheritance on human behavior in populations.

Epigenetic machinery, as a dual information source to a cognitive process, serves as a heritable system, intermediate between (relatively) hard-wired classical genetics, and a (usually) highly Larmarckian embedding cultural context. In particular, the three heritable systems interact, in our model, through a crosstalk in which the epigenetic machinery acts as a kind of intelligent catalyst for gene expression.

4.5 Multiple processes

The argument to this point has, in large measure, been directly abducted from recent formal studies of high level cognition – consciousness – based on a Dretske-style information theoretic treatment of Bernard Baars' global workspace model. A defining and grossly simplifying characteristic of that phenomenon is its rapidity: typically the global broadcasts of consciousness occur in a matter of a few hundred milliseconds, limiting the number of processes that can operate simultaneously. Slower cognitive dynamics can, therefore, be far more complex than individual consciousness. One well known example is institutional distributed cognition that encompasses both individual and group cognition in a hierarchical structure typically operating on timescales ranging from a few seconds or minutes in combat or hunting groups, to years at the level of major governmental structures, commercial enterprises, religious organizations, or other analogous large scale cultural artifacts. Wallace and Fullilove (2008) provides the first formal mathematical analysis of institutional distributed cognition.

Clearly, cognitive gene expression is not generally limited to a few hundred milliseconds, and something much like the distributed cognition analysis may be applied here as well. Extending the analysis requires recognizing an individual cognitive actor can participate in more than one 'task', synchronously, asynchronously, or strictly sequentially. Again, the analogy is with institutional function whereby many individuals often work together on several distinct projects: Envision a multiplicity of possible cognitive gene expression dual 'languages' that themselves form a higher order network linked by crosstalk.

Next, describe crosstalk measures linking different dual languages on that meta-meta (MM) network by some characteristic magnitude ω, and *define a topology on the MM network by renormalizing the network structure to zero if the crosstalk is less than ω and set it equal to one if greater or equal to it.* A particular ω, of sufficient magnitude, defines a giant component of network elements linked by mutual information greater or equal to it, in the sense of Erdos and Renyi, (1960), as more fully described in Wallace and Fullilove (2008, Section 3.4).

The fundamental trick is, in the Morse Theory sense (Matsumoto 2002), to invert the argument so that a given topology for the giant component will, in turn, define some critical value, ω_C, so that network elements interacting by mutual information less than that value will be unable to participate, will be locked out and not active. ω becomes an epigenetically syntactically-dependent detection limit, and depends critically on the instantaneous topology of the giant component defining the interaction between possible gene interaction MM networks.

Suppose, now, that a set of such giant components exists at some generalized system 'time' k and is characterized by a set of parameters $\Omega_k \equiv \omega_1^k, ..., \omega_m^k$. Fixed parameter values define a particular giant component set having a particular set of topological structures. Suppose that, over a sequence of times the set of giant components can be characterized by a possibly coarse-grained path $\gamma_n = \Omega_0, \Omega_1, ..., \Omega_{n-1}$ having significant serial correlations that, in fact, permit definition of an adiabatically, piecewise stationary, ergodic (APSE) information source Γ.

Suppose that a set of (external or internal) epigenetic signals impinging on the set of such giant components can also be characterized by another APSE information source Z that interacts not only with the system of interest globally, but with the tuning parameters of the set of giant components characterized by Γ. Pair the paths (γ_n, z_n) and apply the joint information argument above, generating a splitting criterion between high and low probability sets of pairs of paths. We now have a multiple workspace cognitive genetic expression structure driven by epigenetic catalysis.

4.6 'Coevolutionary' development

The model can be applied to multiple interacting information sources representing simultaneous gene expression processes, for example across a spatially differentiating organism as it develops. This is, in a broad sense, a 'coevolutionary' phenomenon in that the development of one segment may affect that of others.

Most generally, we assume that different cognitive developmental subprocesses of gene expression characterized by information sources H_m interact through chemical or other signals and assume that *different processes become each other's principal environments*, a broadly coevolutionary phenomenon.

We write

$$H_m = H_m(K_1...K_s, ...H_j...), j \neq m,$$

(4.10)

where the K_s represent other relevant parameters.

The dynamics of such a system is driven by a recursive network of stochastic differential equations, similar to those used to study many other highly parallel dynamic structures (e.g., Wymer, 1997).

Letting the K_j and H_m all be represented as parameters Q_j, (with the caveat that H_m not depend on itself), one can define, according to the generalized Onsager development of the Appendix,

$$S^m \equiv H_m - \sum_i Q_i \partial H_m / \partial Q_i$$

to obtain a complicated recursive system of phenomenological 'Onsager relations' stochastic differential equations,

$$dQ_t^j = \sum_i [L_{j,i}(t, ...\partial S^m/\partial Q^i...)dt + \sigma_{j,i}(t, ...\partial S^m/\partial Q^i...)dB_t^i],$$

(4.11)

where, again, for notational simplicity only, we have expressed both the H_j and the external K's in terms of the same symbols Q_j.

m ranges over the H_m and we could allow different kinds of 'noise' dB_t^i, having particular forms of quadratic variation that may, in fact, represent a projection of environmental factors under something like a rate distortion manifold (Wallace and Fullilove, 2008; Wallace and Wallace, 2008).

As usual for such systems, there will be multiple quasi-stable points within a given system's H_m, representing a class of generalized resilience modes accessible via punctuation.

Second, however, there may well be analogs to fragmentation when the system exceeds the critical values of K_c according to the approach of Wallace and Wallace (2008). That is, the K-parameter structure will represent full-scale fragmentation of the entire structure, and not just punctuation within it.

We thus infer two classes of punctuation possible for this kind of structure. There are other possible patterns:

[1] Setting equation 4.11 equal to zero and solving for stationary points again gives attractor states since the noise terms preclude unstable equilibria.

[2] This system may converge to limit cycle or 'strange attractor' behaviors in which the system seems to chase its tail endlessly, e.g., the cycle of climate-driven phenotype changes in persistent temperate region plants.

[3] What is converged to in both cases is not a simple state or limit cycle of states. Rather it is an equivalence class, or set of them, of highly dynamic information sources coupled by mutual interaction through crosstalk. Thus 'stability' in this extended model represents particular patterns of ongoing dynamics rather than some identifiable 'state', although such dynamics may be indexed by a 'stable' set of phenotypes.

Here we become enmeshed in a system of highly recursive phenomenological stochastic differential equations, but at a deeper level than the standard stochastic chemical reaction model (e.g., Zhu et al., 2007), and in a dynamic rather than static manner: the objects of this system are equivalence classes of information sources and their crosstalk, rather than simple final states of a chemical system.

4.7 Multiple models

Wallace and Wallace (2009) argue that consciousness may have undergone the characteristic branching and pruning of evolutionary development, particularly in view of the rapidity of currently surviving conscious mechanisms. According to that study, evolution is littered with polyphyletic parallelisms: many roads lead to functional Romes, and consciousness, as a particular form of high order cognitive process operating in real time, embodies one such example, represented by an equivalence class structure that factors the broad realm of necessary conditions information theoretic realizations of Baars' global workspace model. Many different physiological systems, then, can support rapidly shifting, highly tunable, and even simultaneous assemblages of interacting unconscious cognitive modules. Thus Wallace and Wallace conclude that the variety of possibilities suggests minds today may be only a small surviving fraction of ancient evolutionary radiations – bush phylogenies of consciousness pruned by selection and chance extinction.

Even in the realms of rapid global broadcast inherent to real time cognition, they speculate, following a long tradition, that ancient backbrain structures instantiate rapid emotional responses, while the newer forebrain harbors rapid 'reasoned' responses in animal consciousness. The cooperation and competition of these two rapid phenomena produces, of course, a plethora of systematic behaviors.

Since consciousness is necessarily restricted to realms of a few hundred milliseconds, evolutionary pruning may well have resulted in only a small surviving fraction of previous evolutionary radiations. Processes operating on longer timescales may well be spared such draconian evolutionary selection. That is, the vast spectrum of mathematical models of cognitive gene expression inherent to our analysis here, in the context of development times much longer than a few hundred milliseconds, implies current organisms may simultaneously harbor several, possibly many, quite different cognitive gene expression mechanisms.

It seems likely, then, that, with some generality, slow phenomena, ranging from institutional distributed cognition to cognitive gene expression, permit the operation of very many quite different cognitive processes simultaneously or in rapid succession.

One inference is, then, that gene expression and its epigenetic regulation are, for even very simple organisms, far more complex than individual human consciousness, currently regarded as one of the 'really big' unsolved scientific problems.

Neural network models adapted or abducted from inadequate cognitive studies of a generation ago are unlikely to cleave the Gordian Knot of scientific inference surrounding gene expression.

4.8 Epigenetic focus

The Tuning Theorem analysis of the Appendix permits an inattentional blindness/concentrated focus perspective on the famous computational 'no free lunch' theorem of Wolpert and MacReady (1995, 1997). Following closely the arguments of English (1996), Wolpert and MacReady have established that there exists no generally superior function optimizer. There is no 'free lunch' in the sense that an optimizer 'pays' for superior performance on some functions with inferior performance on others. If the distribution of functions is uniform, then gains and losses balance precisely, and all optimizers have identical average performance. The formal demonstration depends primarily upon a theorem that describes how information is conserved in optimization. This Conservation Lemma states that when an optimizer evaluates points, the posterior joint distribution of values for those points is exactly the prior joint distribution. Put simply, observing the values of a randomly selected function does not change the distribution: An optimizer has to 'pay' for its superiority on one subset of functions with inferiority on the complementary subset.

As English describes, anyone slightly familiar with the evolutionary computing literature recognizes the paper template 'Algorithm X was treated with modification Y to obtain the best known results for problems P_1 and P_2.' Anyone who has tried to find subsequent reports on 'promising' algorithms knows that they are extremely rare. Why should this be?

A claim that an algorithm is the very best for two functions is a claim that it is the very worst, on average, for all but two functions. It is due to the diversity of the benchmark set of test problems that the 'promise' is rarely realized. Boosting performance for one subset of the problems usually detracts from performance for the complement.

English argues that hammers contain information about the distribution of nail-driving problems. Screwdrivers contain information about the distribution of screw-driving problems. Swiss army knives contain information about a broad distribution of survival problems. Swiss army knives do many jobs, but none particularly well. When the many jobs must be done under primitive conditions, Swiss army knives are ideal.

Thus, according to English, the tool literally carries information about the task optimizers are literally tools-an algorithm implemented by a computing device is a physical entity.

Another perspective is that a computed solution is simply the product of the information processing of a problem, and, by a very famous argument, information can never be gained simply by processing. Thus a problem X is transmitted as a message by an information processing channel, Y, a computing device, and recoded as an answer. By the Tuning Theorem argument of the Appendix there will be a channel coding of Y that, when properly tuned, is most efficiently transmitted by the problem. In general, then, the most efficient coding of the transmission channel, that is, the best algorithm turning a problem into a solution, will necessarily be highly problem-specific. Thus there can be no best algorithm for all equivalence classes of problems, although there may well be an optimal algorithm for any given class. The tuning theorem form of the No Free Lunch theorem will apply quite generally to cognitive biological and social structures, as well as to massively parallel machines.

Rate distortion, however, occurs when the problem is collapsed into a smaller, simplified, version and then solved. Then there must be a tradeoff between allowed average distortion and the rate of solution: the retina effect. In a very fundamental sense – particularly for real time systems – rate distortion manifolds present a generalization of the converse of the no free lunch arguments. The neural corollary is known as inattentional blindness (Wallace, 2007).

We are led to suggest that there may well be a considerable set of no free lunch-like conundrums confronting highly parallel real-time structures, including epigenetic control of gene expression, and that they may interact in distinctly complicated ways.

Some of these ways, as the next chapter indicates, produce pathological phenotypes.

5
Developmental disorders

5.1 Network information theory

Let U be an information source representing a systematic embedding environmental 'program' interacting with the process of cognitive gene expression, here defined as a complicated set of information sources having source joint uncertainty $H(Z_1, ..., Z_n)$ that guides the system into a particular equivalence class of desired developmental behaviors and trajectories.

To model the effect of U on development one can, most simply, invoke results from network information theory, (Cover and Thomas, 1991, p. 388). Given three interacting information sources, say Y_1, Y_2, Z, the splitting criterion between high and low probability sets of states, taking Z as the external context, is given by

$$I(Y_1, Y_2|Z) = H(Z) + H(Y_1|Z) + H(Y_2|Z) - H(Y_1, Y_2, Z),$$

where, again, $H(...|...)$ and $H(..., ..., ...)$ represent conditional and joint uncertainties. This generalizes to the relation

$$I(Y_1, ..., Y_n|Z) = H(Z) + \sum_{j=1}^{n} H(Y_j|Z) - H(Y_1, ..., Y_n, Z)$$

Thus the fundamental splitting criterion between low and high probability sets of joint developmental paths becomes

$$I(Z_1, ..., Z_n|U) = H(U) + \sum_{j=1}^{n} H(Z_j|U) - H(Z_1, ..., Z_n, U).$$

(5.1)

R. Wallace, D. Wallace, *Gene Expression and Its Discontents,*
DOI 10.1007/978-1-4419-1482-8_5, © Springer Science+Business Media, LLC 2010

Again, the Z_i represent internal information sources and U that of the embedding environmental context.

The central point is that a one step extension of that system via the results of network information theory allows incorporating the effect of an external environmental 'farmer' in guiding cognitive developmental gene expression.

5.2 Embedding ecosystems as information sources

The principal farmer for a developing organism is the ecosystem in which it is embedded, in a large sense. Summarizing briefly the arguments of Wallace and Wallace, (2008), ecosystems, under appropriate coarse graining, often have reconizable grammar and syntax. For example, the turn-of-the-seasons in a temperate climate, for most natural communities, is remarkably similar from year to year in the sense that the ice melts, migrating birds return, trees bud, flowers and grass grow, plants and animals reproduce, the foliage turns, birds migrate, frost, snow, the rivers freeze, and so on in a predictable manner from year to year.

Suppose, then, that we can coarse grain an ecosystem at time t according to some appropriate partition of the phase space in which each division A_j represents a particular range of numbers for each possible species in the ecosystem, along with associated parameters such as temperature, rainfall, humidity, insolation, and so on. We examine longitudinal paths, statements of the form

$$x(n) = A_0, A_1, ..., A_n$$

defined in terms of some 'natural' time unit characteristic of the system. Then n corresponds to a time unit T, so that $t = T, 2T, ..., nT$. Our interest is in the serial correlation along paths. If $N(n)$ is the number of possible paths of length n that are consistent with the underlying grammar and syntax of the appropriately coarse grained ecosystem, for example, spring leads to summer, autumn, winter, back to spring, etc., but never spring to autumn to summer to winter in a temperate climate.

The essential assumption is that, for appropriate coarse graining, $N(n)$, the number of possible grammatical paths, is much smaller than the total conceivable number of paths, and that, in the limit of large n,

$$H \equiv \lim_{n \to \infty} \frac{\log[N(n)]}{n}$$

both exists and is independent of path.

Not all possible ecosystem coarse grainings are likely to lead to this result, as is sometimes the case with Markov models. Holling (1992) in particular emphasizes that mesoscale ecosystem processes are most likely to entrain dynamics at larger and smaller scales, a process Wallace and Wallace (2008) characterize as *mesoscale resonance*, a generalization of the Baldwin effect. See that reference for details, broadly based on the Tuning Theorem.

5.3 Ecosystems farm organismal development

The environmental and ecosystem farming of development may not always be benign.

Suppose we can operationalize and quantify degrees of both overfocus or inattentional blindness (IAB) and of overall structure or environment distortion (D) in the actions of a highly parallel cognitive epigenetic regulatory system. The essential assumption is that the (internal) dual information source of a cognitive structure that has low levels of both IAB overfocus and structure/environment distortion will tend to be richer than that of one having greater levels. This is shown in figure 5.1a, where H is the source uncertainty dual to internal cognitive process, $X = IAB$, and $Y = D$. Regions of low X, Y, near the origin, have greater source uncertainty than those nearby, so $H(X, Y)$ shows a (relatively gentle) peak at the origin, taken here as simply the product of two error functions.

We are, then, particularly interested in the internal cognitive capacity of the structure itself, as paramatized by degree of overfocus and by the (large scale) distortion between implementation and impact. That capacity, a purely internal quantity, need not be convex in the parameter D, taken as characterizing interaction with an external environment, and thus becomes a context for internal measures. Such measures need not themselves be convex in D.

The generalized Onsager argument, based on the homology between information source uncertainty and free energy, as explained more fully in the Appendix, is shown in figure 5.1b. $S = H(X,Y) - XdH/dX - YdH/dY$, the 'disorder' analog to entropy in a physical system, is graphed on the Z axis against the $X - Y$ plane, assuming a gentle peak in H at the origin. Peaks in S, according to theory, constitute repulsive system barriers, which must be overcome by external forces. In figure 5.1b there are three quasi-stable topological resilience modes, in the sense of Section 1.3, marked as A, B, and C. The A region is locked in to low levels of both overfocus and distortion, as it sits in a pocket. Forcing the system in either direction, that is, increasing either IAB or D, will, initially, be met by homeostatic attempts to return to the resilience state A, according to this model.

If overall distortion becomes severe in spite of homeostatic developmental mechanisms, the system will then jump to the quasi-stable state B, a second pocket. According to the model, however, once that transition takes place, there will be a tendency for the system to remain in a condition of high distortion. That is, the system will become locked-in to a structure with high

$$S=H-XdH/dX-YdS/dY$$

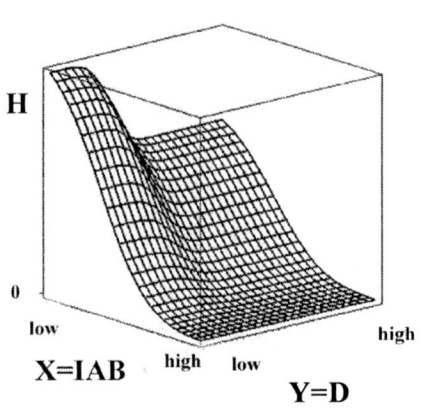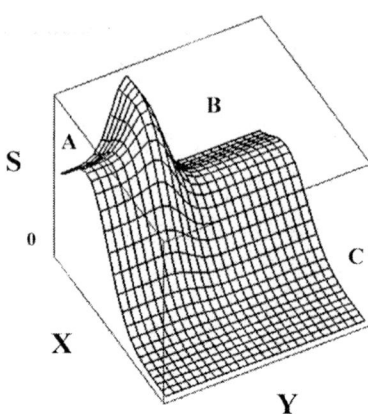

Fig. 5.1. a. Source uncertainty, H, of the dual information source of epigenetic cognition, as parametized by degrees of focus, $X = IAB$ and distortion, $Y = D$, between implementation and actual impact. Note the relatively gentle peak at low values of X, Y. Here H is generated as the product of two error functions. b. Generalized Onsager treatment of figure 5.1a. $S = H(X,Y) - XdH/dX - YdH/dY$. The regions marked A, B, and C represent realms of resilient quasi-stability, divided by barriers defined by the relative peaks in S. Transition among them requires a forcing mechanism. From another perspective, limiting energy or other resources, or imposing stress from the outside, driving down H in figure 5.1a, would force the system into the lower plain of C, in which the system would then become trapped in states having high levels of distortion and inattentional blindness/overfocus.

distortion in the match between structure implementation and structure impact, but one having lower overall cognitive capacity, i.e., a lower value of H in figure 5.1a.

The third pocket, marked C, is a broad plain in which both IAB and D remain high, a highly overfocused, poorly linked pattern of behavior that would require significant intervention to alter once it reached such a quasi-stable resilience mode. The structure's cognitive capacity, measured by H in figure 5.1a, is the lowest of all for this condition of pathological resilience, and attempts to correct the problem – to return to condition A, will be met with very high barriers in S, according to figure 5.1b. That is, mode C is very highly resilient, although pathologically so, much like the eutrophication of a pure lake by sewage outflow. See, e.g., Gunderson (2000) for discussions of ecological resilience and literature references.

We can argue that the three quasi-equilibrium configurations of figure 5.1b represent different dynamical states of the system, and that the possibility of transition between them represents the breaking of the associated symmetry groupoid by external forcing mechanisms. That is, three manifolds representing three different kinds of system dynamics have been patched together to create a more complicated topological structure. For cognitive phenomena, such behavior is likely to be the rule rather than the exception. 'Pure' groupoids are abstractions, and the fundamental questions will involve linkages which break the underlying symmetry.

In all of this, as in equation 4.11, the system convergence is not to some fixed state, limit cycle, or pseudorandom strange attractor, but rather to some appropriate set of highly dynamic information sources, i.e., behavior patterns constituting, here, developmental trajectories, rather than to some fixed 'answer to a computing problem' (Wallace, 2009).

What this model suggests is that sufficiently strong external perturbation can force a highly parallel real-time cognitive epigenetic structure from a normal, almost homeostatic, developmental dynamic into one involving a widespread, comorbid, developmental disorder. This is a well studied pattern for humans and their institutions, reviewed at some length elsewhere (e.g., Wallace and Fullilove, 2008). Indeed, this argument provides the foundation of a fairly comprehensive model of chronic developmental dysfunction across a broad class of cognitive systems, including, but not limited to, cognitive epigenetic control of gene expression. One approach might be as follows:

A developmental process can be viewed as involving a sequence of surfaces like figure 5.1b, having, for example, 'critical periods' when the barriers between the normal state A and the pathological states B and C are relatively low. This might particularly occur under circumstances of rapid growth or long-term energy demand, as discussed in the next section. During such a time the peaks of figure 5.1b might be relatively suppressed, and the system would become highly sensitive to perturbation, and to the onset of a subsequent pathological developmental trajectory.

To reiterate, then, during times of rapid growth, embryonic de- and remethylation, and/or other high system demand, metabolic energy limitation imposes the need to focus via something like a rate distortion manifold. Cognitive process requires energy through the homologies with free energy density, and more focus at one end necessarily implies less at some other. In a distributed zero sum developmental game, as it were, some cognitive or metabolic processes must receive more free energy than others, and these may then be more easily affected by external chemical, biological, or social stressors, or by simple stochastic variation. Something much like this has indeed become a standard perspective (e.g., Waterland and Michels, 2007).

A structure trapped in region C might be said to suffer something much like what Wiegand (2003) describes as the loss of gradient problem, in which one part of a multiple population coevolutionary system comes to dominate the others, creating an impossible situation in which the other participants do not have enough information from which to learn. That is, the cliff just becomes too steep to climb. Wiegand also characterizes focusing problems in which a two-population coevolutionary process becomes overspecialized on the opponent's weaknesses, effectively a kind of inattentional blindness.

Thus there seems some consonance between our asymptotic analysis of cognitive structural function and current studies of pathologies affecting coevolutionary algorithms (e.g., Wallace, 2009; Ficici et al., 2005). In particular the possibility of historic trajectory, of path dependence, in producing individualized failure modes, suggests there can be no one-size-fits-all amelioration strategy.

Equation 5.1 basically enables a kind of environmental catalysis to cognitive gene expression, in a sense closely similar to the arguments of Section 4.1. This is analogous to, but more general than, the 'mesoscale resonance' invoked by Wallace and Wallace (2008): during critical periods, according to these models, environmental signals can have vast impact on developmental trajectory.

5.4 A simple probability argument

Again, critical periods of rapid growth require energy, and by the homology between free energy density and cognitive information source uncertainty, that energy requirement may be in the context of a zero-sum game so that the barriers of figure 5.1 may be lowered by metabolic energy constraints or high energy demand. In particular the groupoid structure of equation 3.1 changes progressively as the organism develops, with new equivalence classes being added to $A = \cup \alpha$. If metabolic energy remains capped, then

$$P[H_\beta] = \frac{\exp[-H_\beta K]}{\sum_\alpha \exp[-H_\alpha K]}$$

must decrease with increase in α, i.e., with increase in the cardinality of A. Thus, for restricted K, barriers between different developmental paths must fall as the system becomes more complicated.

A precis of these results can be more formally captured using methods closely similar to recent algebraic geometry approaches to concurrent, i.e., highly parallel, computing (Goubault and Raussen, 2002; Goubault, 2003; Pratt, 1991).

5.5 Developmental shadows

We now reconsider directed homotopy in a developmental context, as shadowed by critical developmental periods. First, we restrict the analysis to a two dimensional phenotype space, and begin development at some S_0 as in figure 5.2.

If one requires temporal path dependence – no reverse development – then figure 5.2 shows two possible final states, S_1 and S_2, separated by a critical point C *that casts a path-dependent developmental shadow* in time. There are, consequently, two separate 'ways' of reaching a final state in this model. The S_i thus represent (relatively) static phenotypic expressions of the solutions to equation 4.11 that are, of themselves, highly dynamic information sources.

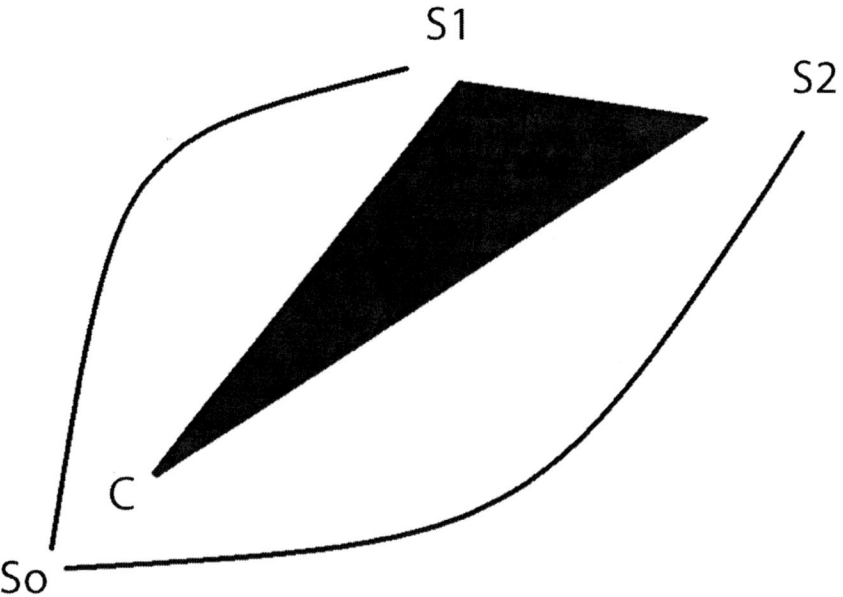

Fig. 5.2. Given an initial developmental state S_0 and a critical period C casting a path-dependent developmental shadow, there are two directed homotopy equivalence classes of deformable paths leading, respectively, to final phenotype states S_1 and S_2 that are expressions of the highly dynamic information source solutions to equation 4.11. These equivalence classes define a topological groupoid on the developmental system.

Elements of each 'way' can be transformed into each other by continuous deformation without crossing the impenetrable shadow cast by the critical period C.

These ways are the equivalence classes defining the system's topological structure, a groupoid analogous to the fundamental homotopy group in spaces that admit of loops (Lee, 2000) rather than time-driven, one-way paths. That is, the closed loops needed for classical homotopy theory are impossible for this kind of system because of the 'flow of time' defining the output of an information source – one goes from S_0 to some final state. The theory is thus one of *directed homotopy*, dihomotopy, and the central question revolves around the continuous deformation of paths in development space into one another, without crossing the shadow cast by the critical period C. Again, Goubault and Raussen (2002) provide another introduction to the formalism.

Thus the external signals U of equation 5.1, as a catalytic mechanism, can define quite different developmental dihomotopies.

Such considerations suggest that a multitasking developmental process that becomes trapped in a particular pattern cannot, in general, expect to emerge from it in the absence of external forcing mechanisms or the stochastic resonance/mutational action of 'noise'. Emerging from such a trap involves large-scale topological changes, and this is the functional equivalent of a first order phase transition in a physical systems and requires energy. It should be clear that the therapeutic forcing of a developing organizm from S_1 to S_2 in figure 5.2 is a major intrusion likely to have a broad spectrum of highly deleterious 'side effects'. Such a shift would represent imposition of a new developmental trajectory, with its own set of possible developmental disorders.

The fundamental topological insight is that environmental context – the U in equation 5.1 – can be imposed on the 'natural' groupoids underlying massively parallel gene expression. This sort of behavior is, as noted above, central to ecosystem resilience theory.

Apparently the set of developmental manifolds, and its subsets of directed homotopy equivalence classes, formally classifies quasi-equilibrium states, and thus characterizes the different possible developmental resilience modes. Some of these may be highly pathological.

Shifts between markedly different topological modes appear to be necessary effects of phase transitions, involving analogs to phase changes in physical systems.

It seems clear that both 'normal development' and possible pathological states can be represented as topological resilience/phase modes in this model, suggesting a real equivalence between difficulties in carrying out gene expression and its stabilization. This mirrors recent results on the relation between programming difficulty and system stability in highly parallel computing devices (Wallace, 2008a).

5.6 Epigenetic programming of artificial systems for biotechnology

Wallace (2009) examines how highly parallel 'Self-X' computing machines – self-programming, protecting, repairing, etc. – are inevitably coevolutionary

5.6 Epigenetic programming of artificial systems for biotechnology

in the sense of Section 4.6 above, since elements of a dynamic structural hierarchy always interact, an effect that will asymptotically dominate system behavior at great scale. The 'farming' paradigm provides a model for programming such devices, that, while broadly similar to the liquid state machines of Maas et al. (2002), differs in that convergence is to an information source, a systematic dynamic, rather than to a computed fixed 'answer'. As the farming metaphor suggests, stabilizing complex coevolutionary mechanisms appears as difficult as programming them. Sufficiently large networks of even the most dimly cognitive modules will become emergently coevolutionary, suggesting the necessity of 'second order' evolutionary programming that generalizes the conventional Nix/Vose models (Wallace, 2008a).

Although we cannot pursue the argument in detail here, very clearly such an approach to programming highly parallel coevolutionary machines – equivalent to deliberate epigenetic farming – should be applicable to a broad class of artificial biological systems/machines for which some particular ongoing behavior is to be required, rather than some final state 'answer'. Examples might include the manufacture, in a large sense, of a dynamic product, e.g., a chemical substance, anti-cancer or artificial immune search-and-destroy strategy, biological signal detection/transduction process, and so on.

Tunable epigenetic catalysis lowers an 'effective energy' associated with the convergence of a highly coevolutionary cognitive system to a final structured dynamic. Given a particular 'farming' information source acting as the program, the behavior of the final state of interest will become associated with the lowest value of the free energy-analog, possibly calculable by optimization methods. If the retina-like rate distortion manifold has been properly implemented, a kind of converse to the no free lunch theorem, then this optimization procedure should converge to an appropriate solution, fixed or dynamic. Thus we invoke a synergism between the focusing theorem and a 'tunable epigenetic catalysis theorem' to raise the probability of an acceptable solution, particularly for a real-time system whose dynamics will be dominated by rate distortion theorem constraints.

The degree of catalysis needed for convergence in a real time system would seem critically dependent on the rate distortion function $R(D)$ or on its product with an acceptable reaction time, τ, that is, on there being sufficient bandwidth in the communication between a cognitive biological 'machine' and its embedding environment. If that bandwidth is too limited, or the available reaction time too short, then the system will inevitably freeze out into what amounts to a highly dysfunctional 'ground state'.

The essential point would seem to be a convergence between emerging needs in biotechnology and general strategies for programming coevolutionary computing devices.

6
An interim perspective

We have, in this approach, hidden the kind of massive calculations made explicit in Ciliberti et al. (2007a, b), burying them as 'fitting regression-model analogs to data', possibly at a second order epigenetic hierarchical level. In the real world such calculations would be quite difficult, particularly given the introduction of punctuated transitions that must be fitted using elaborate renormalization calculations, typically requiring such exotic objects as Lambert W-functions (e.g., Wallace, 2005a).

Analogies with neural network studies suggest, however, intractable conceptual difficulties for spinglass-type models of gene expression and development dynamics, much as claimed by O'Nuallain (2008). In spite of nearly a century of sophisticated neural network model studies – including elegant treatments like Toulouse et al. (1986) – Atmanspacher (2006) claims that to formulate a serious, clear-cut and transparent formal framework for cognitive neuroscience is a challenge comparable to the early stage of physics four centuries ago. Only a very few contemporary approaches, including that of Wallace (2005a), are worth mentioning, in his view.

Furthermore, Krebs (2005) has identified what might well be described as the sufficiency failing of neural network models, that is, neural networks can be constructed as Turing machines that can replicate any known dynamic behavior in the same sense that the Ptolemaic Theory of planetary motion, as a Fourier expansion in epicycles, can, to sufficient order, mimic any observed orbit. Keplerian central motion provides an essential reduction. Krebs' particular characterization is that 'neural possibility is not neural plausibility'.

Likewise, Bennett and Hacker (2003) conclude that neural-centered explanations of high order mental function commit the mereological fallacy, that is, the fundamental logical error of attributing what is in fact a property of an entirety to a limited part of the whole system. 'The brain' does not exist in isolation, but as part of a complete biological individual who is most often deeply embedded in social and cultural contexts.

Neural network-like models of gene expression and development applied to complex living things inherently commit both errors, particularly in a social, cultural, or environmental milieu. This suggests a particular necessity for the formal inclusion of the effects of embedding contexts – epigenetic and environmental – in the sense of Baars (1988, 2005). That is, gene expression and development are conditioned by signals from embedding physiological, social, and for humans, cultural, environments. As described above, our formulation can include such influences in a highly natural manner, as they influence epigenetic catalysis. In addition, multiple, and quite different, cognitive gene expression mechanisms may operate simultaneously, or in appropriate sequence, given sufficient development time.

Developmental disorders, in a broad sense that must include comorbid mental and physical characteristics, emerge as pathological 'resilience' modes, a viewpoint from ecosystem theory quite similar to that of epigenetic epidemiology. Environmental farming through an embedding information source affecting internal epigenetic regulation of gene expression, can, as a kind of programming of a highly parallel cognitive system, place the organism into a quasi-stable pathological developmental pattern converging on a dysfunctional phenotype.

Altering such a phenotype with potent medical interventions – 'epigenetic drugs' – will not necessarily return the organism to health, but trigger a new developmental path that, given the extraordinary nature of the perturbation, converges on a phenotype having a shortened lifespan, an unpleasant, but likely inevitable, 'side effect'.

The probability models of cognitive process presented here will lead to statistical tools based on the asymptotic limit theorems of information theory, in the same sense that the usual parametric statistics are based on the Central Limit Theorem. We have not, then, given 'a' model of development and its disorders, but, rather, outlined a possible general strategy for fitting empirically-determined statistical models to real data, in precisely the sense that one would fit the usual parametric statistical models to normally distributed data. Fitting statistical models does not, of itself, perform scientific inference. That is done through comparing fitted models for similar systems under different, or different systems under similar, conditions, and by examining the structure of residuals.

One implication of this work, then, is that understanding complicated processes of gene expression and development – and their pathologies – will require construction of data analysis tools considerably more sophisticated than the current crop of simple models abducted from neural network studies or stochastic chemical reaction theory. Most centrally, however, presently popular (and fundable) reductionist approaches to understanding gene expression must eventually exhaust themselves in the same desert of sand-grain hyperparticularity that appears to have driven Francis Crick from molecular biology to the study of consciousness, a field now mature enough to provide tools for use in the other direction.

The next chapters apply our perspectives to a number of chronic diseases, emphasizing the role of psychosocial stressors in their etiology. It will become increasingly apparent that, while our model of epigenetic catalysis applied to developmental disorders works well at the individual or simple aggregate levels of analysis, understanding community patterns of chronic disease and their dynamics will require a significant extension of theory. Human ecosystems are no less complex than their 'natural' counterparts, particularly given the central role of culture in human biology (e.g., Richerson and Boyd, 2004; Schultz, 2009). Thus individual scale analyses must be extended to incorporate both cross-scale linkages and emergent mechanisms of collective gene expression and cultural lock-in of epigenetic mechanism that transcend individual organismal response to environmental signals. This should not be unexpected, as the individual cells of a multicellular organism must collectively coordinate their own gene expression patterns and dynamics in response to that organism's environment.

Ultimately, human populations host three inheritance systems, i.e., cultural, epigenetic, and genetic, sometimes acting on widely different time scales, and coming iterations of mathematical theory must embrace their synergisms.

7
The obesity pandemic in the US

7.1 Introduction

Obesity is epidemic in the United States, has been so for more than two decades, and continues to increase. The condition is rapidly becoming pandemic worldwide (e.g., Egger and Swinburn, 1997; Kimm and Obarzanek, 2002; Roth et al., 2004). Current rates of overweight in the US are 61% and 14% in adults and children respectively. Obesity in adults has nearly doubled since 1980, from 15% in 1980 to 27% by 1999 (Welman and Friedberg, 2002). Childhood overweight is rapidly rising in the US, particularly among African Americans and Hispanics. By 1998 prevalence increased to 21.5% in African Americans, 21.8% among Hispanics, and 12.3% among non-Hispanic whites aged 4 to 12 years (Strauss and Pollack, 2001).

The obesity pandemic is associated with serious health conditions including type 2 diabetes, heart disease, high blood pressure and stroke, certain types of cancer, hypoxia, sleep apnea, hernia, and arthritis. It is a major cause of economic loss, death, and suffering that shows no indications of abatement.

A series of articles in *Science* (Vol. 299, 7 February, 2003) focused on an individualized perspective that largely fails to explore the epidemiology and population ecology of the problem. The piece by Hill et al. (2003), for example, explains the obesity epidemic as the simple disjunction between calorie intake and output consequent on eased workload, 'larger food portions', and disinclinations to exercise. In contrast, our perspective places the 'explanation' that 'obesity occurs when people eat too much and get too little exercise' in the same category as the remark by US President Calvin Coolidge on the eve of the Great Depression that 'unemployment occurs when large numbers of people are out of work'. Both statements ignore profound structural determinants of great population suffering that must be addressed by collective actions of equally great reform.

Indeed, experts on health disparities have long recognized that obesity is unevenly distributed geographically, ethnically, and by socioeconomic class. Urban people of color (Allen, 1998), poor Southern states (Mokdad et al.,

1999), and poor neighborhoods within cities (Ginsberg-Fellner et al., 1981) have higher prevalences. The Southern states form the epicenter of the geographically spreading epidemic (Mokdad et al., 1999), a picture of contagion between populations.

The famous Whitehall Studies of British civil servants (Brunner et al., 1997) found that coronary heart disease and central abdominal fat deposition incidences were strongly associated with the occupational hierarchy. Locus of work control was a major factor in both central abdominal fat deposition and coronary heart disease. Power relations in the workplace imposed a particular structure of stress.

Furthermore, stress which causes sleep deficits shifts metabolism toward fat accumulation and central abdominal deposition (Spiegel et al., 1999). The Hypothalamic-Pituitary-Adrenal (HPA) axis is central to the mechanisms (Bjorntorp, 2001; Chrousos, 2000). So the stress involves adrenal reactions to serious threats.

Our hypothesis, in contrast to that of the *Science* special issue, is that large numbers of Americans feel seriously threatened. The obesity epidemic embodies the consequences of public policies: economic insecurity from deindustrialization, social upheaval from destruction of cities by programs of 'benign neglect' and 'planned shrinkage', (Wallace and Wallace, 1998), the nation's wealth increasingly concentrated in fewer and fewer hands, and a voting ritual that, as the 2000 US presidential election implied, doesn't seem to matter. The current political opening to the contrary, the US obesity epidemic embodies, in our view, a worsening crisis of democratic locus-of-control that will not be addressed by platitudes about 'eating less and exercising more'.

7.2 Stress and the HPA axis

Abdominal obesity and visceral fat accumulation are particularly associated with disease, and have become the focus of much research on 'stress' and its relation to the 'fight-or-flight' responses of the HPA axis. Section 1.4 briefly paraphrased Bjorntorp (2001), who extensively summarized the role of the HPA axis in physiological responses to stress. In sum, stress-driven increased activity of the HPA axis triggers inhibition of both the pituitary gonadal and growth hormone axes. Stress may, then, synergistically cause accumulation of visceral fat, via elevated cortisol secretion and decrease of sex steroid and growth hormones. Bjorntorp concludes that the deposit of central body fat, closely correlated with general measures of obesity, can serve as a reasonable approximation to long-term endocrine abnormalities associated with stress and often-repeated or chronic activation of the HPA axis.

That is, stress literally writes an image of itself onto the body as visceral fat accumulation, first having written an image of itself onto the HPA axis. The phenomenon can be interpreted as the transmission of a structured signal between communicating systems, in a large sense, i.e., from psychosocial structure to HPA. In the sense of the Rate Distortion Theorem of the Mathematical Appendix the structured stressors are the y^n of Section 14.4 and the response of the HPA axis are the b^n, that can be retranslated back into a distorted version of the stressors, $\hat{y}^n$, as in the theorem. The $\hat{y}^n$ would then be the actual distorted image of the structured stressors, represented in shorthand by the HPA signals b^n. We will develop this argument below.

We look at how the communication of the embedding psychosocial structure and the HPA axis might be constrained by certain of the asymptotic limit theorems of probability, using the perspective of the earlier chapters.

Stress is not often random in human societies, nor is it undifferentiated like pressure under water. It is most often highly structured, in essence a 'language', having both a grammar and a syntax, so that certain stressors are 'meaningful' in a particular context, and others are not, having little or no long-term physiological effect (e.g., Miller and Chen, 2006; Clougherty et al., 2007). We will argue that the HPA axis is, in fact, a cognitive system itself associated with a dual information source that may also be expressed as a kind of language. The punctuated interaction of these two 'languages' is critical to an understanding of how psychosocial stress affects the HPA axis, and, ultimately, writes a distorted image of itself on the human body as visceral fat deposition.

This analysis presents a slightly different picture of the obesity epidemic, but one having profound implications for intervention policy.

7.3 HPA axis cognition

Atlan and Cohen (1998) argue that the essence of cognition is comparison of a perceived external signal with an internal picture of the world, and then, upon that comparison, the choice of a response from a much larger repertoire of possible responses. Clearly, from this perspective, the HPA axis, the 'flight-or-fight' reflex, is cognitive. Upon recognition of a new perturbation in the surrounding environment, emotional and/or conscious cognition evaluate and choose from several possible responses: no action necessary, flight, fight, helplessness (flight or fight needed, but not possible). Upon appropriate conditioning, the HPA axis is able to accelerate the decision process, much as the immune system has a more efficient response to second pathogenic challenge once the initial infection has become encoded in immune memory. Certainly hyperreactivity as a sequela of post traumatic stress disorder (PTSD) is a well known example.

7.4 Interacting information sources

In the usual manner we suppose that the behavior of the HPA axis can be represented by a sequence of states in time, the path $x \equiv x_0, x_1, ...$ Similarly, we assume an external signal of structured psychosocial stress can be similarly represented by a path $y \equiv y_0, y_1, ...$ These paths are, however, both very highly structured and, within themselves, are serially correlated and can, in fact, be represented by information sources $\mathbf{X}$ and $\mathbf{Y}$. We assume the HPA axis and the external stressors interact, so that these sequences of states are not independent, but are jointly serially correlated. We can, then, define a path of sequential pairs as $z \equiv (x_0, y_0), (x_1, y_1), ...$

The essential content of the Joint Asymptotic Equipartition Theorem (JAEPT) is that the set of joint paths z can be partitioned into a relatively small set of high probability which is termed *jointly typical*, and a much larger set of vanishingly small probability. Further, according to the JAEPT, the splitting criterion between high and low probability sets of pairs is the mutual information

$$I(X,Y) = H(X) - H(X|Y) = H(X) + H(Y) - H(X,Y)$$

(7.1)

where $H(X), H(Y), H(X|Y)$ and $H(X,Y)$ are, respectively, the Shannon uncertainties of X and Y, their conditional uncertainty, and their joint uncertainty.

To recapitulate, suppose that these two (adiabatically piecewise stationary) ergodic information sources $\mathbf{Y}$ and $\mathbf{B}$ begin to interact, to 'talk' to each other, i.e., to influence each other in some way so that it is possible, for example, to look at the output of $\mathbf{B}$ – strings b – and infer something about the behavior of $\mathbf{Y}$ from it – strings y. We suppose it possible to define a retranslation from the B-language into the Y-language through a deterministic code book, and call $\hat{\mathbf{Y}}$ the translated information source, as mirrored by $\mathbf{B}$.

Define some distortion measure comparing paths y to paths $\hat{y}$, $d(y, \hat{y})$, following the arguments of the Mathematical Appendix. We invoke the Rate Distortion Theorem's mutual information $I(Y, \hat{Y})$, which is the splitting criterion between high and low probability pairs of paths.

Extending the analyses, triplets of sequences can be divided by a splitting criterion into two sets, having high and low probabilities respectively. For large n the number of triplet sequences in the high probability set will be determined by the relation (Cover and Thomas, 1991, p. 387]

$$N(n) \propto \exp[nI(Y_1; Y_2|Y_3)],$$

(7.2)

where splitting criterion is given by

$$I(Y_1; Y_2|Y_3) \equiv$$

$$H(Y_3) + H(Y_1|Y_3) + H(Y_2|Y_3) - H(Y_1, Y_2, Y_3)$$

We can then examine mixed cognitive/adaptive symmetry changes amounting to phase transitions analogous to learning plateaus in the splitting criterion $I(Y_1, Y_2|Y_3)$. Again these results are roughly analogous to the Eldredge/Gould model of evolutionary punctuated equilibrium (Gould, 2002).

Note, again, that the expression above can be generalized to a number of interacting information sources, Y_j, embedded in a larger context, Z, as

$$I(Y_1, ..., Y_s|Z) = H(Z) + \sum_{j=1}^{s} H(Y_j|Z) - H(Y_1, ..., Y_s, Z)$$

7.5 The simplest HPA axis model

Stress, as we envision it, is not a random sequence of perturbations, and is not independent of its perception. Rather, it involves a highly correlated, grammatical, syntactical process by which an embedding psychosocial environment communicates with an individual, particularly with that individual's HPA axis, in the context of social hierarchy. We view the stress experienced by an individual as an adiabatically piecewise stationary ergodic (APSE) information source, interacting with a similar dual information source defined by HPA axis cognition.

Again, the ergodic nature of the language of stress is essentially a generalization of the law of large numbers, so that long-time averages can be well approximated by cross-sectional expectations. Languages do not have simple autocorrelation patterns, in distinct contrast with the usual assumption of random perturbations by white noise in the standard formulation of stochastic differential equations.

Let us suppose we cannot measure stress, but can determine the concentrations of HPA axis hormones and other biochemicals according to some natural time frame, that we will characterize as the inherent period of the system. Suppose, in the absence of extraordinary meaningful psychosocial stress, we measure a series of n concentrations at time t which we represent as an n-dimensional vector X_t. Suppose we conduct a number of experiments, and create a regression model so that we can, in the absence of perturbation, write, to first order, the concentration of biomarkers at time $t+1$ in terms of that at time t using a matrix equation of the form

$$X_{t+1} \approx <\mathbf{R}> X_t + b_0,$$

(7.3)

where $<\mathbf{R}>$ is the matrix of regression coefficients and b_0 a vector of constant terms.

We then suppose that, in the presence of a perturbation by structured stress

$$X_{t+1} = (<\mathbf{R}> + \delta \mathbf{R}_{t+1})X_t + b_0$$

$$\equiv <\mathbf{R}> X_t + \epsilon_{t+1},$$

(7.4)

where we have absorbed both b_0 and $\delta \mathbf{R}_{t+1}X_t$ into a vector ϵ_{t+1} of error terms that are not necessarily small in this formulation. In addition it is important to realize that this is not a population process whose continuous analog is exponential growth. Rather what we examine is more akin to the passage of a signal – structured psychosocial stress – through a distorting physiological filter.

If the matrix of regression coefficients $<\mathbf{R}>$ is sufficiently regular, we can (Jordan block) diagonalize it using the matrix of its column eigenvectors $\mathbf{Q}$, writing

$$\mathbf{Q}X_{t+1} = (\mathbf{Q} <\mathbf{R}> \mathbf{Q}^{-1})\mathbf{Q}X_t + \mathbf{Q}\epsilon_{t+1},$$

(7.5)

or equivalently as

7.5 The simplest HPA axis model

$$Y_{t+1} = <\mathbf{J}> Y_t + W_{t+1},$$

(7.6)

where $Y_t \equiv \mathbf{Q}X_t, W_{t+1} \equiv \mathbf{Q}\epsilon_{t+1}$, and $<\mathbf{J}> \equiv \mathbf{Q}<\mathbf{R}>\mathbf{Q}^{-1}$ is a (block) diagonal matrix in terms of the eigenvalues of $<\mathbf{R}>$.

Thus the (rate distorted) writing of structured stress on the HPA axis through $\delta\mathbf{R}_{t+1}$ is reexpressed in terms of the vector W_{t+1}.

The sequence of W_{t+1} is the rate-distorted image of the information source defined by the system of external structured psychosocial stress. This formulation permits estimation of the long-term steady-state effects of that image on the HPA axis. The essential trick is to recognize that because everything is APSE, we can either time or ensemble average both sides of equation (7.6), so that the one-period offset is absorbed in the averaging, giving an 'equilibrium' relation

$$<Y> = <\mathbf{J}><Y> + <W>$$

or

$$<Y> = (\mathbf{I} - <\mathbf{J}>)^{-1} <W>,$$

(7.7)

where $\mathbf{I}$ is the $n \times n$ identity matrix.

Now we reverse the argument: Suppose that Y_k is chosen to be some fixed eigenvector of $<\mathbf{R}>$. Using the diagonalization of $<\mathbf{J}>$ in terms of its eigenvalues, we obtain the average excitation of the HPA axis in terms of some eigentransformed pattern of exciting perturbations as

$$<Y_k> = \frac{1}{1 - <\lambda_k>} <W_k>$$

(7.8)

where $<\lambda_k>$ is the eigenvalue of $<Y_k>$, and $<W_k>$ is some appropriately transformed set of ongoing perturbations by structured psychosocial stress.

The essence of this result is that *there will be a characteristic form of perturbation by structured psychosocial stress – the W_k – that will resonantly excite a particular eigenmode of the HPA axis.* Conversely, by tuning the eigenmodes of $<\mathbf{R}>$, the HPA axis can be trained to galvanized response in the presence of particular forms of perturbation.

This is because, if $<\mathbf{R}>$ has been appropriately determined from regression relations, then the λ_k will be a kind of multiple correlation coefficient (Wallace and Wallace, 2000), so that particular eigenpatterns of perturbation will have greatly amplified impact on the behavior of the HPA axis. If $\lambda = 0$ then perturbation has no more effect than its own magnitude. If, however, $\lambda \to 1$, then the written image of a perturbing psychosocial stressor will have very great effect on the HPA axis. Following Ives (1995), we call a system with $\lambda \approx 0$ *resilient* since its response is no greater than the perturbation itself.

We suggest, then, that learning by the HPA axis is, in fact, the process of tuning response to perturbation. This is why we have written $<\mathbf{R}>$ instead of simply $\mathbf{R}$: The regression matrix is a tunable set of variables.

Suppose we require that $<\lambda>$ itself be a function of the magnitude of excitation, i.e.,

$$<\lambda> = f(|<W>|),$$

where $|<W>|$ is the vector length of $<W>$. We can, for example, require the amplification factor $1/(1-<\lambda>)$ to have a signal transduction form, an inverted-U-shaped curve, for example the signal-to-noise ratio of a stochastic resonance, so that

$$\frac{1}{1-<\lambda>} = \frac{1/|<W>|^2}{1+b\exp[1/(2|<W>|)]}.$$

(7.9)

This places particular constraints on the behavior of the learned average $<\mathbf{R}>$, and gives precisely the typical HPA axis pattern of initial hypersensitization, followed by anergy or burnout with increasing average stress, a behavior that might well be characterized as pathological resilience.

Variants of this model permit imposition of cycles of different length, for example hormonal on top of circadian. Typically this is done by requiring a cyclic structure in matrix multiplication, with a new matrix $\mathbf{S}$ defined in terms of a sequential set of the $\mathbf{R}$, having period m, so that

$$\mathbf{S}_t \equiv \mathbf{R}_{t+m}\mathbf{R}_{t+m-1}...\mathbf{R}_t.$$

Essentially one does matrix algebra 'modulo m', in a sense.

In general, while the eigenvalues of such a cyclic system may remain the same, its eigenvectors depend on the choice of phase, i.e., where you start in the cycle. This is a complexity of no small note, and could represent a source of contrast in HPA axis behavior between men and women, beyond that driven by the ten-fold difference in testosterone levels.

7.6 Obesity as a developmental disorder

Hirsch (2002) suggests that obesity is a developmental disorder with roots in utero or early childhood. He and others have developed a set point or homeostatic theory of body weight, finding that it is the process that determines that set point which needs examination, rather than the homeostasis itself, which is now fairly well understood. Hirsch concludes that the truly relevant question is not why obese people fail treatment, it is how their level of fat storage became elevated, a matter, he concludes, is probably rooted in infancy and childhood, when strong genetic determinants are shaping a still-plastic organism.

The question we raised earlier regarding the division of sets of possible responses of a cognitive HPA axis into the sets B_0 and B_1 of Section 2.2 has special significance in this matter.

Recall that the essential characteristic of cognition in our formalism involves a function $h(x)$ that maps a mixed algorithmic path $x = a_0, a_1, ..., a_n, ...$ onto a member of one of two disjoint sets, B_0 or B_1. Thus respectively, either (1) $h(x) \in B_0$, implying no action taken, or (2), $h(x) \in B_1$, and some particular response is chosen from a large repertoire of possible responses. We suggest here that some higher order cognitive module might be needed to identify what constitutes B_0, the set of 'normal' states. This is because, in the absence of epigenetic catalysis, there is no low energy mode for information systems. Virtually all states are more or less high energy states involving high rates of information transfer, and there is no way to identify a ground state using the physicist's favorite variational or other extremal arguments. An analog to epigenetic catalysis via an interacting infomation source provides such a mechanism.

Suppose that higher order cognitive module, which we might well characterize as a Zero Mode Identification (ZMI) catalytic mechanism, interacts with an embedding language of structured psychosocial stress and, instantiating a Rate Distortion image of that embedding stress, begins to include one or more members of the set B_1 into the set B_0. If that element of B_1 involves a particular mode of HPA axis cortisol/leptin cycle interaction, then the bodymass set point may be reset to a higher value: onset of obesity. This is another form of epigenetic catalysis, although not in a genetic context.

Work by Barker and colleagues suggests that those who develop coronary heart disease (CHD) grow differently from others, both in utero and during childhood. Slow growth during fetal life and infancy is followed by accelerated weight gain in childhood, setting a life history trajectory for CHD, type II diabetes, hypertension, and, of course, obesity. Barker (2002) concludes that slow fetal growth might also heighten the body's stress responses and increase vulnerability to poor living conditions later in life. Thus faulty ZMI function at critical periods in growth could lead to a permanently high body mass set point as a developmental disorder.

Empirical tests of our higher order hypothesis, however, quickly lead into real-world regression models involving the interrelations of measurable biomarkers, behaviors, and so on, requiring formalism much like that used in the section above.

7.7 Recent trajectories of structured stress in the US

Two powerful and intertwining phenomena of socioeconomic disintegration – deurbanization in the 1970's, and deindustrialization, particularly since 1980 – have combined to profoundly damage many US communities, dispersing historic accumulations of economic, political, and social capital. These losses have had manifold and persisting impacts on both institutions and individuals (Pappas, 1989; Ullmann, 1988; Wallace and Wallace, 1998). Elsewhere we examined the effect of these policy-driven phenomena on the hierarchical diffusion of AIDS in the US (Wallace et al., 1999). Here we extend that work to their association with obesity, in the context of the causal biological model given above.

By 1980, not a single African-American urban community established before or during World War II remained intact. Many Hispanic urban neighborhoods established after the war suffered similar fates. Virtually all lost considerable housing, population, and economic and social capital either to programs of urban renewal in the 1950's or to policy-related contagious urban decay from the late 1960's through the late 1970's (Fullilove, 2004; Wallace and Wallace, 1998; Wallace and Fullilove, 2008).

Figure 7.1 exemplifies the process, showing the percent change of occupied housing units in the Bronx section of New York City between 1970 and 1980 by Health Area, the aggregation of US Census Tracts by which morbidity and mortality are reported in the city. The South-Central section of the Bronx, by itself one of the largest urban concentrations in the Western world with about 1.4 million inhabitants, had lost between 55 and 80 percent of housing units, most within a five year period. This is a level of damage unprecedented in an industrialized nation short of civil or international war, and indeed can be construed as constituting a kind of covert civil war (Duryea, 1978; Wallace and Wallace, 1998).

Figure 7.2, a composite index of number and seriousness of building fires from 1959 through 1990 (Wallace and Wallace, 1998; Wallace et al., 1999), illustrates the process of contagious urban decay in New York City producing that housing loss, affecting large sections of Harlem in Manhattan, and

7.7 Recent trajectories of structured stress in the US   83

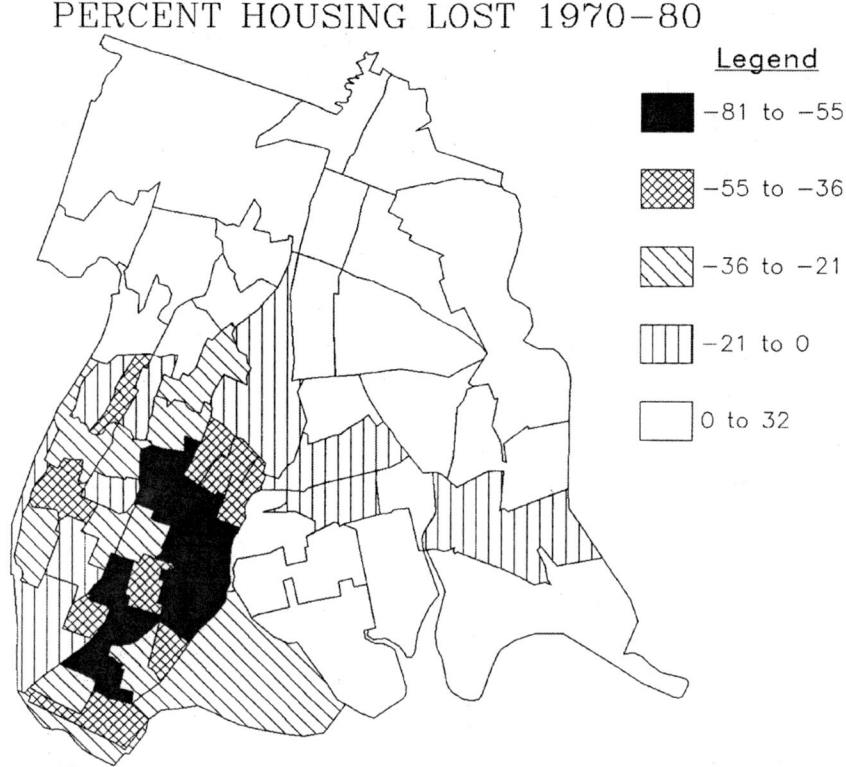

Fig. 7.1. Percent change in occupied housing units, Bronx section of New York City, 1970-1980. Large areas lost over half their housing in this period, a degree of destruction unprecedented in an industrialized nation outside of wartime. Similar policy-driven disasters have afflicted most US urban minority communities since the end of World War II.

a broad band across the African-American and Hispanic neighborhoods of Northern Brooklyn, from Williamsburg to Bushwick, Brownsville, and East New York. The sudden rise between 1967 and 1968 was stemmed through 1972 by the opening of 20 new fire companies in high fire incidence, minority neighborhoods of the city. Beginning in late 1972, however, some 50 firefighting units were closed and many others destaffed as part of a 'planned shrinkage' program that continued the ethnic cleansing policies of 1950's urban renewal without benefit of either constitutional niceties or new housing construction to shelter the displaced population (Fullilove, 2004; Wallace and Wallace, 1998; Wallace and Fullilove, 2008).

Similar maps and graphs could be drawn for devastated sections of Detroit, Chicago, Los Angeles, Philadelphia, Baltimore, Cleveland, Pittsburgh,

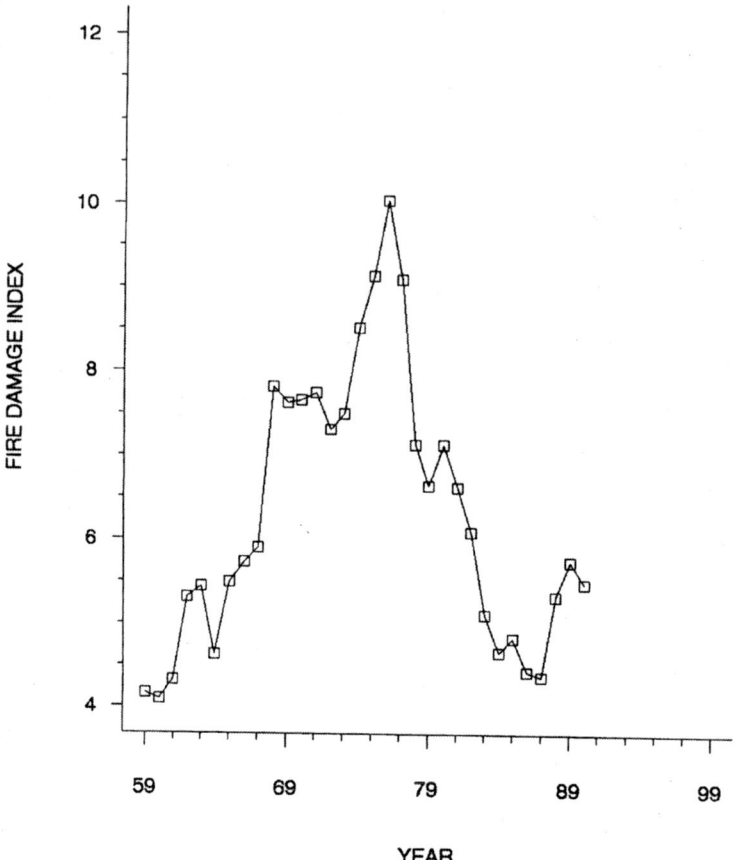

Fig. 7.2. Annual fire damage index, New York City, 1959-1990, a composite of number and seriousness of structural fires, and an index of contagious urban decay. Some 20 new fire companies were added to high fire areas between 1969 and 1971, interrupting the process. Fifty firefighting units were closed in or permanently relocated from, high fire areas after November, 1972, allowing contagious urban decay to proceed to completion, producing the conditions of figure 7.1.

7.7 Recent trajectories of structured stress in the US

Newark, and a plethora of smaller US urban centers, each with its own individual story of active public policy and passive 'benign neglect'.

Figure 7.1 represents the Bronx part of the spatial distribution of the time integral of figure 7.2.

Figure 7.3, using data taken from the US Census, shows the counties of the Northeastern US losing more than 1000 manufacturing jobs between 1972 and 1987, the famous rust belt. It is, in its way, an exact parallel to figure 7.1 in that unionized manufacturing jobs lost remained lost, and their associated social capital and political influence were dispersed. As Pappas (1989) describes, the effects were profound and permanent:

> By 1982 mass unemployment had reemerged as a major social issue [in the USA]. Unemployment rose to its highest level since before World War II, and an estimated 12 million people were out of work – 10.8 percent of the labor force in the nation. It was not, however, a really new phenomenon. After 1968 a pattern was established in which each recession was followed by higher levels of unemployment during recovery. During the depth of the 1975 recession, national unemployment rose to 9.2 percent. In 1983, when a recovery was proclaimed, unemployment remained at 9.5 percent annually.
>
> Certain sectors of the work force have been more heavily affected than others. There was a 16.9 percent jobless rate among blue-collar workers in April, 1982... Unemployment and underemployment have become major problems for the working class. While monthly unemployment figures rise and fall, these underlying problems have persisted over a long period. Mild recoveries merely distract out attention from them.

Figure 7.4, using data from the US Bureau of Labor Statistics, shows the total number of US manufacturing jobs from 1980 to 2001. We define our environmental index of the US national pattern of structured stress to be represented by *the integral of manufacturing job loss after 1980*, i.e., the space between the observed curve and a horizontal line drawn out from the 1980 number of jobs. This is not quite the same as figure 7.3, that represents a simple net loss between two time periods. We believe that manufacturing job loss at one period continues to have influence at subsequent periods as a consequence of permanently dispersed social and political capital, at least over a 20 year span.

Other models, perhaps with different integral weighting functions, are, of course, possible. We use simply

$$D(T) = - \sum_{\tau=1980}^{\tau=T} [M(\tau) - M(1980)]$$

86   7 The obesity pandemic in the US

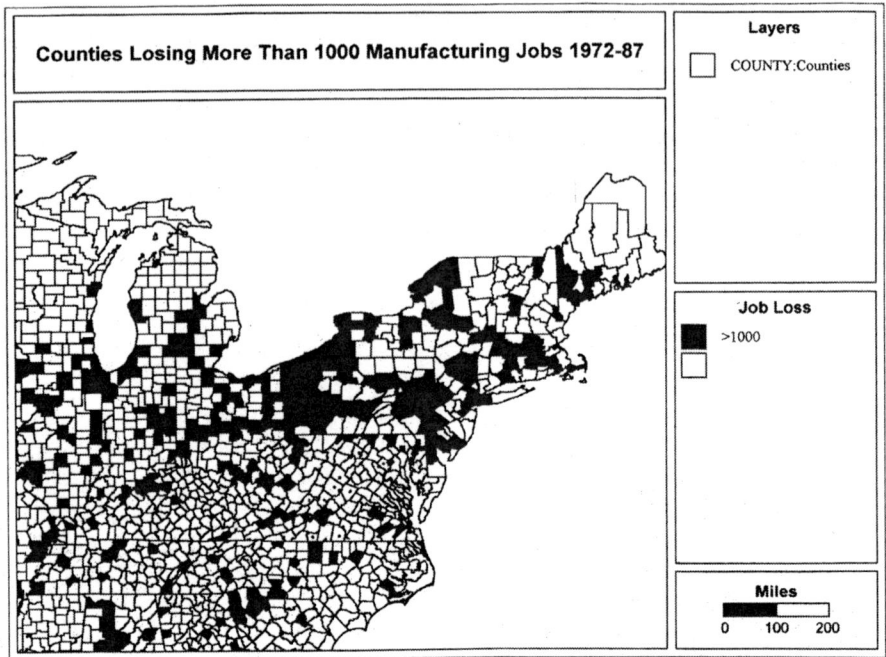

Fig. 7.3. The Rust Belt: Counties of the Northeastern US which lost 1000 or more manufacturing jobs between 1972 and 1987.

(7.10)

while a more elaborate treatment might involve something like

$$D(T) = -\int_{\tau_0}^{T} f(T-\tau)M(\tau)d\tau$$

(7.11)

where D is the deficit, $M(\tau)$ is the number of manufacturing jobs at time τ, and $f(T-\tau)$ is a lagged weighting function.

Figure 7.5, using data from the Centers for Disease Control (2003), shows the percent of US adults characterized as obese according to the Behavioral Risk Factor Surveillance System between 1991 and 2001. This is given as a

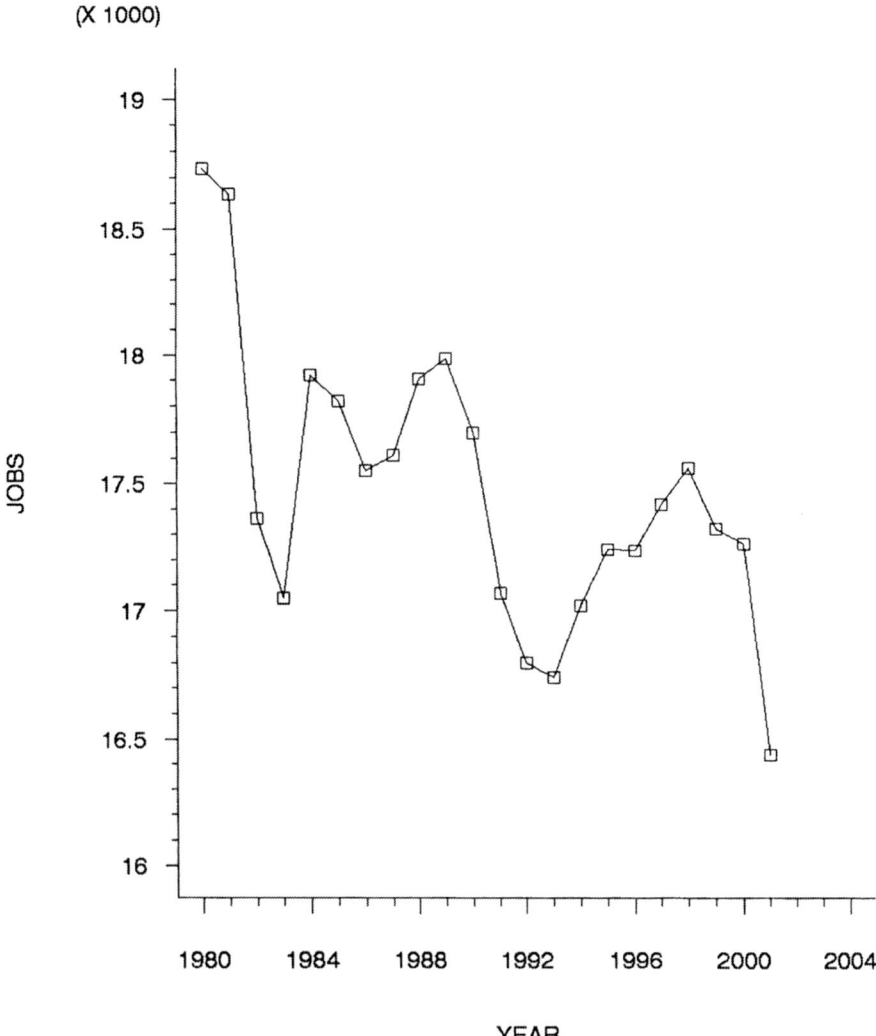

Fig. 7.4. Annual number of manufacturing jobs in the US, 1980-2001. Our environmental index of social decay is the integrated loss after the 1980 peak, representing the permanent dispersal of economic, social, and political capital, part of the opportunity cost of a deindustrialization largely driven by the diversion of technical resources from civilian industry into the Cold War (e.g., Ullmann, 1988).

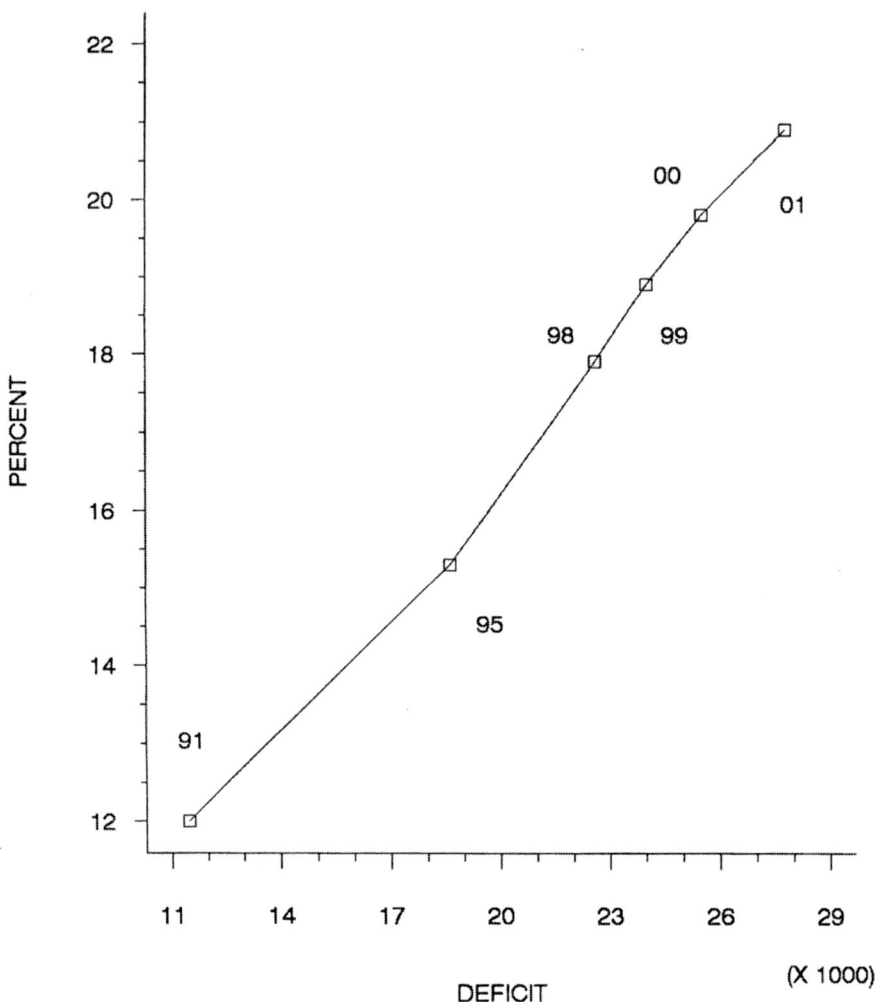

Fig. 7.5. 1991-2001 relation between adult obesity in the US and the integrated loss of manufacturing jobs after 1980. We believe manufacturing job loss is an index of permanent decline in social, economic, and political capital which is perceived as, and indeed represents, a serious threat to the well-being of the US population.

function of the integrated manufacturing jobs deficit from 1980, again, calculated as a simple negative sum of annual differences from 1980.

The association is quite good indeed, and the theory of the first sections suggests the relation is causal and not simply correlational: Loss of stable working class employment, loss of social and political capital, loss of union influence on working conditions and public policies, deurbanization intertwined with deindustrialization and their political outfalls, all constitute a massive threat expressing itself in population-level patterns of HPA axis-driven metabolism and metabolic syndrome.

Figure 7.6 extends the analysis to diabetes deaths in the US between 1980 and 1998. It shows the death rate per 100,000 as a function of the cumulative manufacturing jobs deficit from 1980 through 1998. Diabetes deaths are, after a lag, a good index of population obesity. Two systems are evident, before and after 1989, with a phase transition between them probably representing, in Holling's (1973) sense, a change in ecological domain roughly analogous to the sudden eutrophication of a lake progressively subjected to contaminated runoff. This would seem to reflect the delayed cumulative impacts of both deindustrialization and the deurbanization which became closely coupled with it. Further sudden, marked, upward transitions seem likely if socioeconomic and political reforms are not forthcoming. A simple linear correlation for the period gives an R^2 of 0.91, not inconsiderable.

It would be useful to compare annual county-level maps of diabetes death rates with those of manufacturing job loss and deurbanization, but such a study would require considerable resources in order to conduct the necessary sophisticated analyses of cross-coupled, lagged, spatial and social diffusion.

Figure 7.7 shows a regression of the Black vs. White diabetes death rates (per 100,000) in the US for the period 1979-1997. It is striking that, while the rate of increase for African-Americans was more than 50 % higher than for whites, both subpopulations were closely linked together in a relentless progression: The R^2 of the regression was 0.99. Similarly, figure 7.8 shows Black vs. White hypertension death rates (per 100,000) over the same time span. Again, while African-Americans suffered proportionally more than Whites, the two groups were closely linked in a remarkable joint increase, $R^2 = 0.85$.

Diabetes and hypertension are, of course, both closely related to obesity.

7.8 Confronting the obesity epidemic

Current theory clearly identifies stress as critical to the etiology of visceral obesity, the metabolic syndrome, and their pathological sequelae, mediated by the HPA axis and several other physiological subsystems which we have not addressed here.

Both animal and human studies, however, have indicated that not all stressors are equal in their effect: particular forms of domination in animals and

7 The obesity pandemic in the US

DIABETES DEATH RATE VS. CUMULATIVE MANUFACTURING JOBS DEFICIT 1980-98

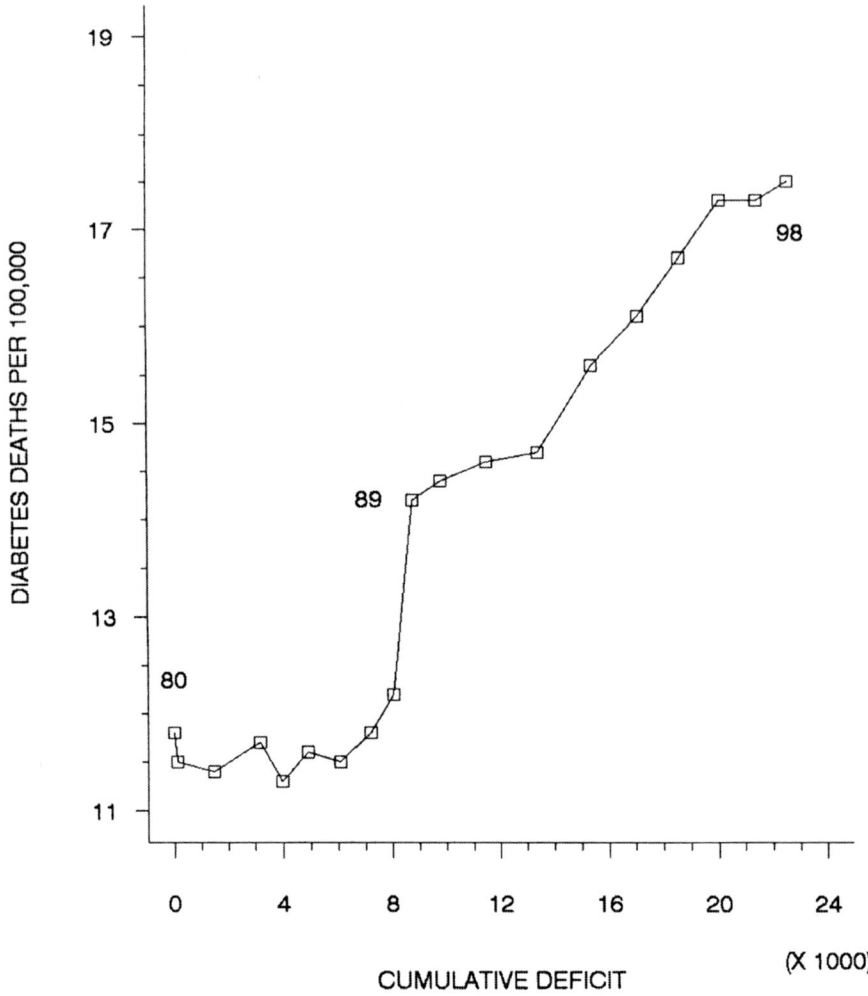

Fig. 7.6. 1980-1998 relation between US diabetes death rate and integrated loss of manufacturing jobs after 1980. Two systems are evident: before and after 1989. We believe this sudden change represents a nonlinear transition between ecosystem domains which is much like the eutrophication of an increasingly contaminated water body (e.g., Holling, 1973). The simple correlation has $R^2 = 0.91$.

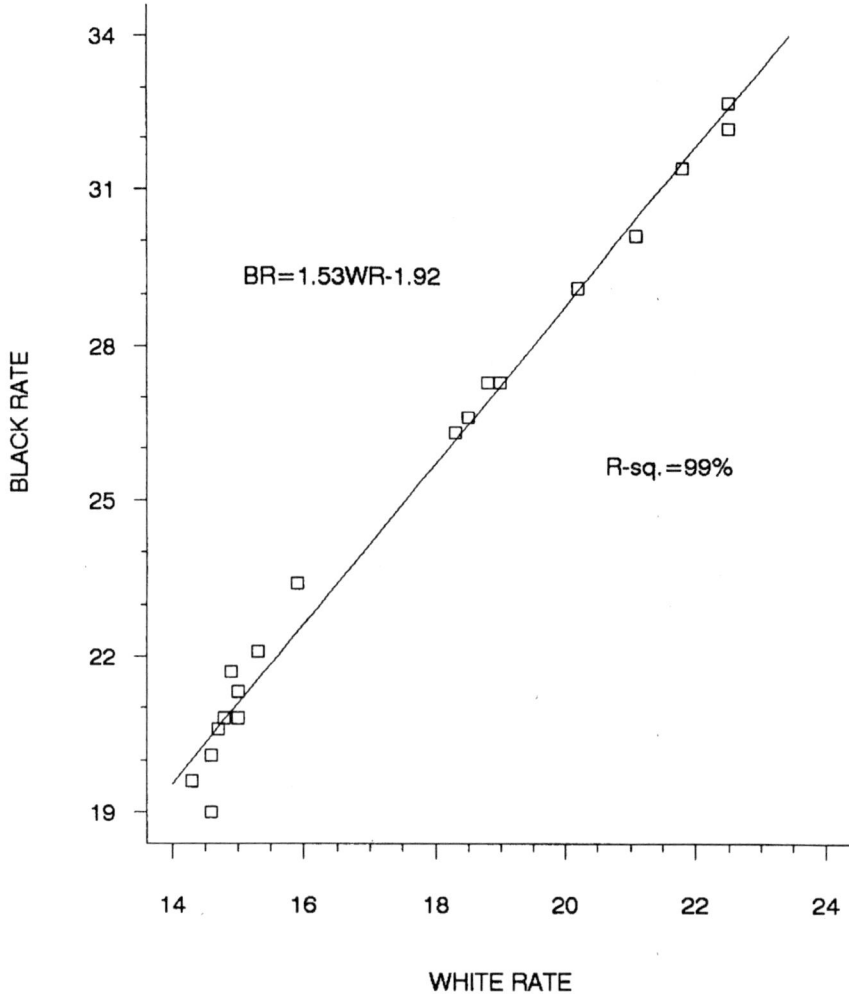

Fig. 7.7. US Black vs. White diabetes death rates (per 10^5), 1979-97. While the Black rates are uniformly higher, strong coupling implies both populations are closely linked in a deteriorating social structure.

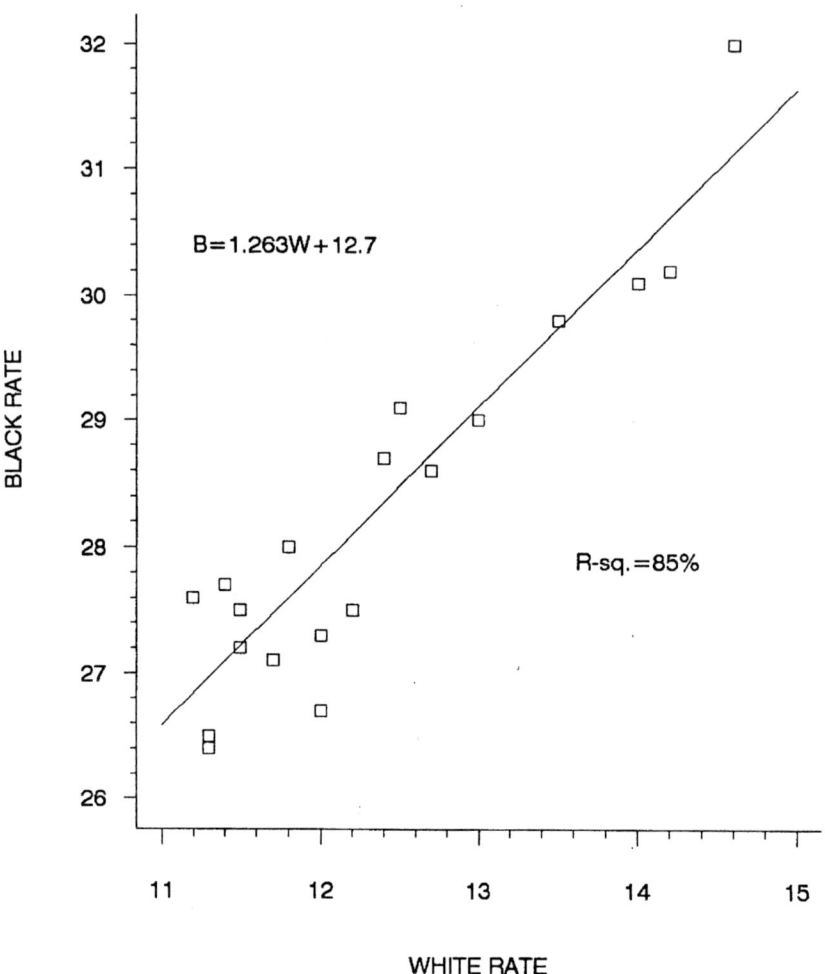

Fig. 7.8. Same as figure 7.7 for US hypertension death rates.

lack of control over work activities in humans are well-known to be especially effective in triggering metabolic syndrome and chronic inflammatory coronary lipid deposition. Thus stress can be given meaning from context, in O'Nuallain's sense.

Recent analyses have examined the general association between social status and health in Western subcultures. For example, figure 7.9, from Singe-Manoux et al. (2003), displays a clear dose-response relation between age adjusted prevalence of self-reported ill-health versus self-reported status rank for white collar workers in the UK. 1 is high and 10 low rank. The low status group approaches the 'LD-50' level at which half the population shows a response to dosage.

For the US, figure 7.10 shows the percent of income concentrated in the top five percent of the population as a function of the integral of manufacturing job loss, 1980-1998. The correlation is very high indeed, suggesting that the destruction of the US industrial base consequent on the catastrophic diversion of scientific and engineering resources into the Cold War (e.g., Wallace, et al., 1999) has had the effect of concentrating wealth and power in the hands of a very small segment of the population. The loss of unionized industrial jobs, and their guarantees of job security, health insurance, retirement benefits, and the exercise of collective power is, in our view, a principal source of a widening population level stress.

Figure 7.11 shows the simple linear correlation between the annual percent of the US voting age population convicted of a felony between 1980 and 1998 and the integral of manufacturing job loss for the period. The correlation is very good indeed. The percent of felons tripled, serving as yet another index of, and significant contributor to, population level stress.

Our analysis has been in terms of a cognitive HPA axis responding to a highly structured 'language' of psychosocial stress, that we see as literally writing a distorted image of itself onto the behavior of the HPA axis in a manner analogous to learning plateaus in a neural network or to punctuated equilibrium in a simple evolutionary process. The first form of phase transition/generalized symmetry change might be regarded as representing the progression of a normally 'staged' disease. The other could describe certain pathologies characterized by stasis or only slight change, with staging a rare (and perhaps fatal) event. The works of Barker and his group suggests that such HPA axis dysfunction in a mother can become a strong epigenetic catalyst for her children.

Psychosocial stress is, for humans, a cultural artifact, one of many such that interact intimately with human physiology. Indeed, much current theory in evolutionary anthropology focuses on the essential (but not unique) role culture plays in human biology (Avital and Jablonka, 2000; Durham, 1991).

If, as the evolutionary anthropologist Robert Boyd has suggested, 'Culture is as much a part of human biology as the enamel on our teeth', what does the rising tide of obesity in the US suggest about American culture and the American system? About 22% of both African-American and Hispanic chil-

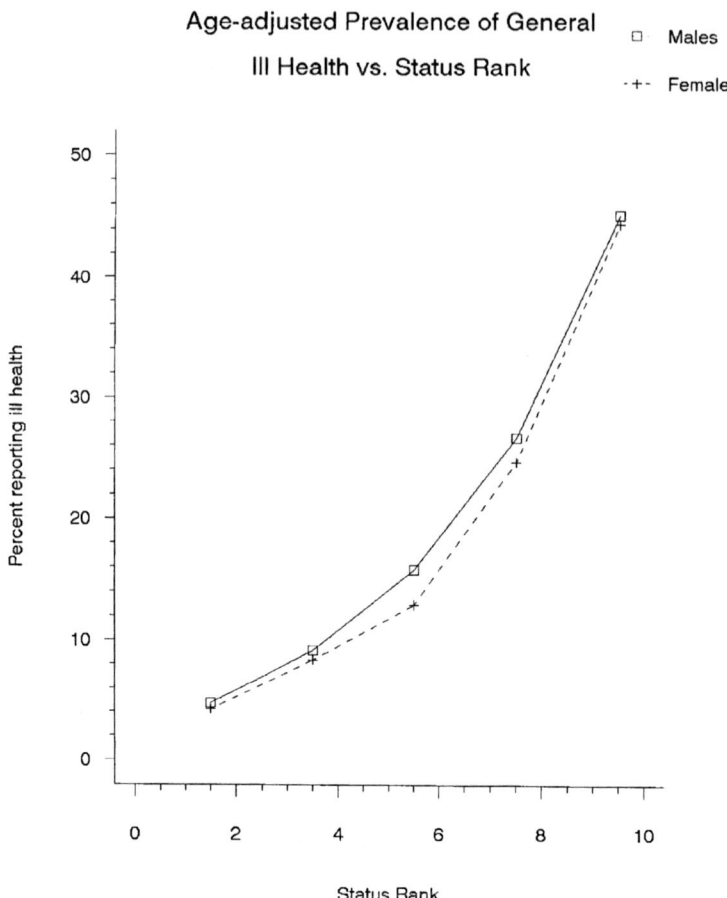

Fig. 7.9. Redisplay of data from Singh-Manoux et al. (2003). Sex-specific dose-response curves of age-adjusted self-reported ill-health vs. self-reported status rank, Whitehall II cohort, 1997 and 1999. 1 is high, 10 is low, status. The curve is approaching the LD-50 at which half the dosed population suffers physiological effect of a poison.

7.8 Confronting the obesity epidemic   95

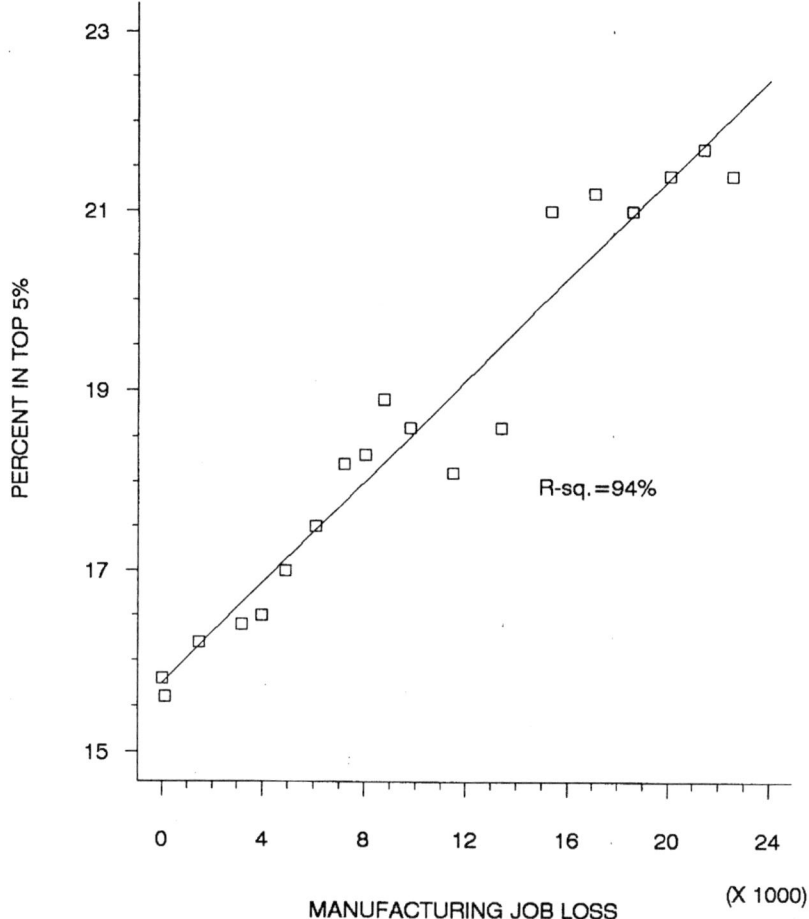

Fig. 7.10. Percent of total US income concentrated in the top 5 percent as a function of the integral of manufacturing job loss, 1980-1998. Devastation of the manufacturing sector consequent on the disruptions caused by the Cold War diversion of engineering and scientific resources from the civilian economy has disempowered vast sections of the US population.

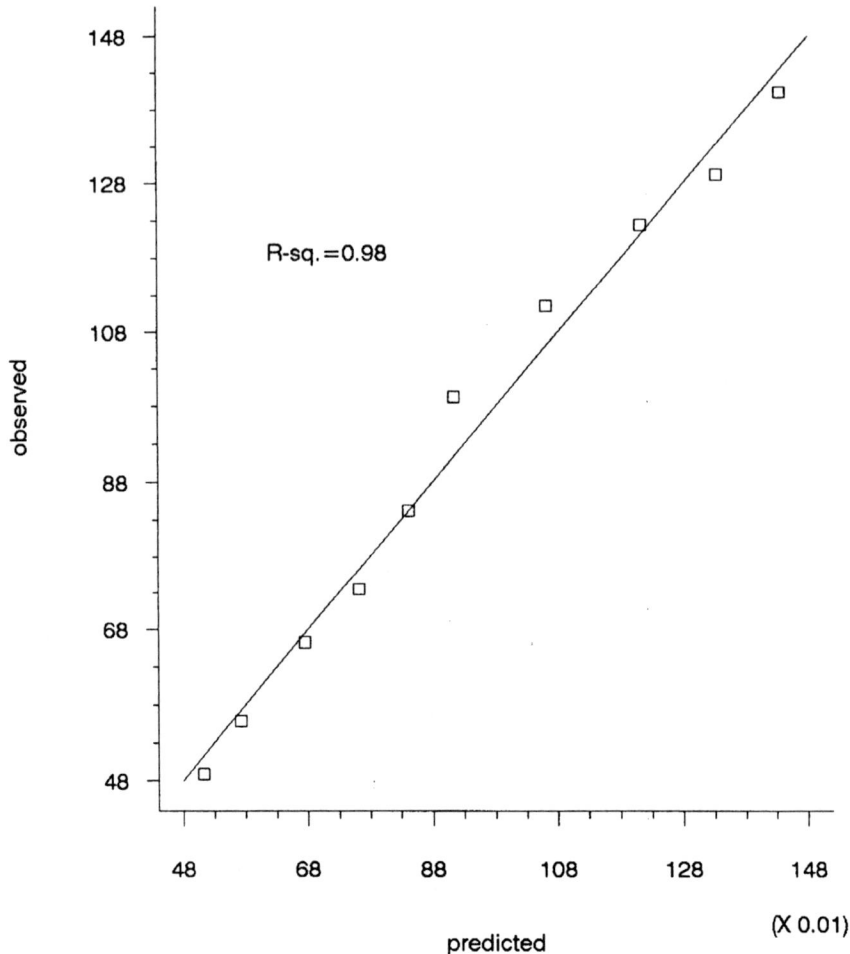

Fig. 7.11. Felons as a percent of the US voting-age population, 1980-1998, expressed as a linear function of the integral of the manufacturing job loss over the period. The proportion of felons tripled. The high correlation suggests that the loss of stable, unionized, manufacturing jobs in the US affects public order as well as public health, and these deteriorations, of course, are likely to be powerfully synergistic.

dren are overweight, as compared to about 12% of non-Hispanic whites, and that prevalence is rising across the board (Strauss and Pollack, 2001). This suggests that, while the effects of an accelerating social pathology related to deindustrialization, deurbanization, and loss of democracy may be most severe for ethnic minorities in the US, the larger, embedding, cultural dysfunction has already spread upward along the social hierarchy, and is quickly entraining the majority population as well.

This is an explanation whose policy implications stand in stark contrast to current individual-oriented exhortations about 'taking responsibility for one's behavior' or 'eating less and getting more exercise' (Hill et al., 2003). The US liberal approach is to mirror the explanations of the failed drug war: People overeat because there's a McDonald's on every street corner, companies market bigger portions, the food they sell is fatty, and so on. In contrast, we find that the fundamental cause of the obesity epidemic over the last twenty years is not television, the automobile, or junk food. All were significant features of American life from the late 1950's into the 1980's without an obesity epidemic. The fundamental cause of the US obesity epidemic is a massive threat to the population caused by continuing deterioration of basic US social, economic, and related structures, in the particular context of a ratcheting of dominance relations resulting from the concentration of effective power within a shrinking elite. This phenomenon is literally writing a life-threatening image of itself onto the bodies of American adults and children. There is already a large and growing literature on other aspects of the sharpening inequalities within the US system (particularly Wilkinson, 1996, and related material), and our conclusions fit within that body of work. Such inequalities imply, as well, increasing actual deprivation that we will find, in Chapters 11 and 12, appears to be the principal driving force at the community scale, where synergistic mesoscale interactions may overwhelm the individual and simple aggregate levels of analysis at which our theory seems best to apply.

The basic and highly plieotropic nature of the biological relation between structured psychosocial stress and cognitive physiological systems ensures that 'magic bullet' interventions will be largely circumvented: in the presence of a continuing socioeconomic and political ratchet, 'medical' modalities are likely to provide little more than the equivalent of a choice of dying by hanging or by firing squad.

Effective intervention against obesity in the US is predicated on creation of a broad, multi-level, ecological control program. It is evident that such a program must include redress of the power relations between groups, rebuilding of urban (and, increasingly, suburban) minority communities, and effective reindustrialization. This implies the necessity of a resurgence of the labor union, religious, civil rights, and community-based political activities which have been traditionally directed against cultural patterns of injustice in the past, activities which, ultimately, liberate all.

8
Coronary heart disease in the US

8.1 Introduction

The origin of 'racial', class, and ethnic disparities in health has recently become the center of some debate in the US, with remedies proposed by mainstream authorities characteristically and predictably focused on individual-oriented 'prevention' by altered life-style or related medical magic bullet interventions. See Link and Phelan (2000) for a comprehensive critique and review. Indeed, for certain US subpopulations, changes in diet, exercise, patterns of smoking and alcohol intake, and so forth, are widely credited with causing markedly declining incidence of coronary heart disease (CHD): the national death rate from CHD for white US males declined from about 420 per 100,000 in 1980 to about 240 by 1997, compared to declines from 350 to 235 for Black males (Cooper, 2000, 2001). Declines for both Black and white females have not been as spectacular, starting from a lower 1980 baseline of about 240, and falling to near 150 by 1997.

The declines have not been uniformly distributed: CHD mortality rates in the US are especially high in middle-aged black men relative to other race/sex groups. Barnett and Halverson (2000) found unexpectedly high rates of premature CHD mortality for African Americans in major metropolitan regions outside the South, despite favorable levels of socioeconomic resources. Fang et al. (1998), find close correlation of CHD mortality with patterns of racial segregation in New York City, one of the world's most segregated urban centers. More generally, similar work (Polednak, 1991, 1993, 1996; Collins and Williams, 1996) shows that all-cause black-white mortality differences are highest in US metropolitan areas with the greatest racial segregation.

Kiecolt-Glaser et al. (2002) discuss how chronic inflammation has recently been linked with a spectrum of conditions associated with aging, including cardiovascular disease (CVD), osteoporosis, arthritis, type II diabetes, certain cancers, and other conditions. The association between CVD and inflammation is mediated by the cytokine IL-6, related to its central role in promoting the production of C-reactive protein (CRP), an ancient and highly conserved protein secreted by the liver in response to trauma, inflammation, and infection (DuClos, 2000; Volanakis, 2001). CRP is a pattern-recognition molecule of the innate immune response keyed to surveillance for altered self and certain pathogens, thus providing early defense and activation of the humoral, adaptive, immune system. It is increasingly seen as a linkage between the two forms of immune response.

Inflammation has recently become central to understanding the etiology of CHD, including its staging as a chronic disease. To paraphrase Blake and Ridker (2001), from initial stages of leukocyte recruitment to diseased endothelium, to plaque rupture, inflammatory mechanisms mediate key steps in atherogenesis and its complications. Thus the key to CHD is now seen in the complex basic biology of plaque formation and dynamics rather than in a passive and rather bland lipid storage.

Triggers for inflammation in atherogenesis include hypertension, diabetes, and obesity. Obesity not only predisposes to insulin resistance and diabetes, but also contributes to atherogenic dyslipidemia; obesity itself promotes inflammation and potentiates atherogenesis independent of effects on insulin resistance or lipoproteins (Libbey et al., 2001).

Libbey et al.(2001) also conclude that atherothrombosis is more than a disease of lipid accumulation, rather it is a disorder characterized by low-grade vascular inflammation, often associated with traditional risk factors such as central obesity and high body mass index. Data implicate inflammatory pathways in all stages of disease, from early atherogenesis, to the progression of lesions, and finally in the thrombotic complications of the disease.

As Ridker (2002) states, risk factors for arteriosclerosis and adult-onset diabetes closely overlap and the two disorders may derive from similar antecedents, a mutual inflammatory or genetic basis. Baseline levels of IL-6 and CRP which were previously shown to predict onset of atherothrombosis, also predict onset of type II diabetes, even after adjustment for body mass index.

Significantly, Ridker also notes the limits of such a strict biochemical approach, finding that the clinical hypothesis that an enhanced immune response results in increased plaque vulnerability begs the question as to why a population distribution of inflammation exists in the first place and what the underlying determinants of this distribution might be.

This question is, precisely, the principal focus of our analysis, seen through the prism of our earlier arguments on epigenetic catalysis.

Indeed, larger hypotheses are not lacking in the literature, and the well-recognized cortisol-leptin cycle is worthy of some comment here: Leptin, the 'fat hormone', increases Th1 and suppresses Th2 cytokine production (Lord et al., 1998) and also stimulates proliferation and activation of circulating monocytes, and may play a direct role in inflammatory processes (Santos-Alvarez et al., 1999). Leptin and cortisol have, however, a complex relation. Cortisol, an adrenal stress hormone, and leptin alternate their plasma peaks as part of the normal circadian cycle (Bornstein et al., 1998). Cortisol increases can trigger answering leptin increases (Newcomer et al., 1988). Glucocorticoid levels also influence plasma leptin levels (Eliman et al, 1998). Thus leptin and the adrenal hormones regulate each other: patterns of stress thus influence weight change, disease resistance, and inflammatory response. Th1/Th2 balance may be heavily influenced, in turn, by the adrenal hormone/leptin balance. Stress imposed on pregnant women may result in changed fetal immune and metabolic processes, with implications for birth weight, fat metabolism and risk for cardiovascular disease and allergenic susceptibility over the life course.

These inferences are strengthened by the results of Singhal et al. (2002) who found that elevation in leptin was associated with impaired vascular function independent of metabolic and inflammatory disturbances associated with obesity.

A long series of articles by Barker and co-workers (Barker, 2002; Barker et al., 2002; Osmond et al., 2000; Godfrey and Barker, 2001) is consistent with such mechanisms, suggesting that those who develop CHD grow differently from others both in utero and during childhood. Slow growth during fetal life and infancy is followed by accelerated weight gain in childhood, setting a life history trajectory for CHD, type II diabetes, and hypertension. Barker and colleagues conclude that slow fetal growth might also heighten the body's stress responses and increase vulnerability to poor living conditions in later life. Thus, in this view, CHD is a developmental disorder that originates through two widespread biological phenomena, developmental plasticity and compensatory growth, a speculation consistent with the work of Smith et al. (1998) who found that deprivation in childhood influences risk of mortality from CHD in adulthood, although an additive influence of adulthood circumstances is seen in such cases.

These latter results are important for incorporating a second body of research, regarding a particular kind of work stress on the development of CHD. Kivimaki et al. (2002) have found that high job strain and 'effort-reward imbalance' increase the risk of cardiovascular mortality, reinforcing a large body of work showing the adverse effect of a wage-slavery work environment on CHD (e.g., Bosma et al., 1997, 1998, 2001). Both effort-reward imbalance and lack of control over one's job seem to contribute to development of adult CHD.

Figure 7.9 above redisplays material on hierarchy and health in a recent paper by Singh-Manoux et al. (2003). To reiterate, that figure is taken from Phase V of the Whitehall II study of London-based office staff, aged 35-55, working in 20 Civil Service departments in 1997 and 1999. The data, covering about 7000 men and 3400 women, show the age-adjusted percent reporting ill-health as a function of self-reported status rank, where 1 is high and 10 is low. Self-reported health is a highly significant predictor of both morbidity and mortality. The results are quite remarkable, both for their evident nonlinearity and for the high prevalence of somaticized distress among the lower-ranked staff. As Link and Phelan argue (2000), rank in this case would seem to inextricably convolute lack of job control with significant de-facto deprivation, for example effective exposure to new health information.

Figure 7.9 is, we claim, very precisely the initial part of a classic S-shaped dose-response curve for exposure of a population to a physiologically-active substance. This one has very nearly reached the fifty percent effective concentration level, or EC-50, for the sample, that is, the dosage at which half of the exposed individuals display the observed physiological response, here age-adjusted self-reported illness. The pattern is highly consistent with assertions that social conditions – in this case a particular form of hierarchy – in fact represent social exposures synergistic with other physiologically active agents, for example classic toxic substances. The analysis is, however, complicated by the essential role culture in human life (Richerson and Boyd, 2004).

CHD seems, then, to be very much a life-history disease associated with a particular kind of sociocultural environment – what we call pathogenic hierarchy. We shall be interested in a model of how such an environment might write itself onto immune function. Our argument, a straightforward adaptation of recent developments in evolutionary theory, couples fundamental human biological mechanisms across multiple scales.

We begin with a restatement of the emerging theory of immune function as immune cognition, and explore linkage with both central nervous system (CNS) cognition, and with the cognitive processes of an embedding sociocultural network. We will then subject this multiply-synergistic cognitive process to patterns of externally-imposed 'structured stress'. Using a rate distortion approach, we find such pressure can literally write a distorted version of itself downward in scale onto immune function as chronic inflammation. Thus the special role of culture in human biology (e.g., Singh-Manoux et al., 2003; Durham, 1991; Richerson and Boyd, 2004), particularly as associated with social hierarchy, becomes directly and organically manifest in the basic biology and dynamics of plaque formation.

That is, for human populations cultural factors like racism, wage slavery, and exaggerated social disparity, are as much a part of the basic biology of coronary heart disease as are the molecular or biochemical mechanisms of plaque deposition and development, via mechanisms of, or akin to, epigenetic catalysis.

In essence we will argue that pathogenic social hierarchy constitutes a directly active force that writes an image of itself upon immune function as chronic vascular inflammation and its sequelae.

8.2 Cognition, immune cognition, and culture

Interactions between the central nervous system (CNS) and the immune system, and between the genetic heritage and the immune system, have become officially recognized and academically codified through journals with titles such as Neuroimmunology and Immunogenetics. Here we will argue that a cognitive socioculture – a social network embodying culture – in which individuals are embedded, and through which they are both acculturated and function to meet collective challenges of threat and opportunity, may interact strongly with individual immune function to produce a composite entity that might well be labeled an Immunocultural Condensation (ICC).

We examine current visions of the interaction between genes and culture, and between the CNS and culture, and follow with a summary of Cohen's view of immune cognition. Next we argue that immune cognition and cognitive socioculture can become fused into a composite entity – the ICC – and that this composite, in turn, can be profoundly influenced by embedding systems of highly structured psychosocial and socioeconomic stressors via mechanisms analogous to epigenetic catalysis. In particular, we argue that the internal structure of the external stress – its 'grammar' and 'syntax' – are important in defining the coupling with the ICC.

Early work by Wallace and colleagues (Wallace 2002a, b, Wallace et al., 2003) presented a detailed mathematical model of the ICC and its linkage with structured patterns of psychosocial or socioeconomic stress. The necessity of such an approach emerges from examination of the theory of immune cognition, although more recent analysis, in the chapters above, views these influences as forms of epigenetic catalysis.

Increasingly, biologists are roundly excoriating simple genetic reductionism which neglects the role of environment. Lewontin (2000), for example, explains that genomes are not 'blueprints,' a favorite public relations metaphor, as genes do not 'encode' for phenotypes. Organisms are instead outgrowths of fluid, conditional interactions between genes and their environments, as well as developmental noise. Organisms, in turn, shape their environments, generating what Lewontin terms a triple helix of cause and effect. Such interpenetration of causal factors may be embodied by an array of organismal phenomena, including, as we shall discuss, culture's relationships with the brain and the immune system. We propose reinterpreting immune function in this light, in particular the coupling of the individual immune system with larger, embedding structures.

The current vision of human biology among evolutionary anthropologists is consistent with Lewontin's analysis and is summarized by Durham (1991) as follows:

> ...[G]enes and culture constitute two distinct but interacting systems of inheritance within human populations... [and] information of both kinds has influence, actual or potential, over ... behaviors [which] creates a real and unambiguous symmetry between genes and phenotypes on the one hand, and culture and phenotypes on the other...
>
> [G]enes and culture are best represented as two parallel lines or 'tracks' of hereditary influence on phenotypes...

With regard to such melding, over hominid evolution genes came to encode for increasing hypersociality, learning, and language skills, so the complex cultural structures which better aid in buffering the local environment became widespread in successful populations (Bonner, 1980).

Every successful human population seems to have a core of tool usage, sophisticated language, oral tradition, mythology and music, focused on relatively small family/extended family groupings of various forms. More complex social structures are build on the periphery of this basic genetic/cultural object (Wallace and Wallace, 1999).

At the level of the individual human, the genetic-cultural object appears to be mediated by what evolutionary psychologists postulate are cognitive modules within the human mind (Barkow et al., 1992). Each module was shaped by natural selection in response to specific environmental and social conundrums Pleistocene hunter-gatherers faced. One set of such domain-specific cognitive adaptations addresses problems of social interchange (e.g., Cosmides and Tooby, 1992). The human species' very identity may rest, in part, on its unique evolved capacities for social mediation and cultural transmission.

Indeed, a brain-and-culture condensation has been adopted as a kind of new orthodoxy in recent studies of human cognition. For example Nisbett et al. (2001) review an extensive literature on empirical studies of basic cognitive differences between individuals raised in what they call 'East Asian' and 'Western' cultural heritages. They view Western-based pattern cognition as 'analytic' and East-Asian as 'holistic,' finding that:

1. Social organization directs attention to some aspects of the perceptual field at the expense of others.

2. What is attended to influences metaphysics.

3. Metaphysics guides tacit epistemology, that is, beliefs about the nature of the world and causality.

4. Epistemology dictates the development and application of some cognitive processes at the expense of others.

5. Social organization can directly affect the plausibility of metaphysical assumptions, such as whether causality should be regarded as residing in the field vs. in the object.

6. Social organization and social practices can directly influence the development and use of cognitive processes such as dialectical vs. logical ones.

Nisbett et al. conclude that tools of thought embody a culture's intellectual history, that tools have theories build into them, and that users accept these theories, albeit unknowingly, when they use these tools.

We may assume, then, the existence of gene-culture and brain-culture condensations.

As described above, Atlan and Cohen (1998) proposed an information-theoretic adaptation of Cohen's (1991, 2000) cognitive principle model of immune function and process, a paradigm incorporating pattern recognition behaviors analogous to those of the central nervous system. Their work follows a long tradition of similar 'cognitive' hypotheses regarding immune function, particularly comparison of the immune system's elaborate chemical network with the brain's neural network, an approach which was well expressed in Jerne's 1967 Nobel Prize talk (e.g., Grossman, 1989, 1992a, b, 1993, 2000).

The reader may wish to review the first part of Section 1.3.2 for details of the Atlan/Cohen model.

As shown at length in the mathematical formalism of the earlier chapters, it is possible to give Atlan and Cohen's language metaphor of meaning-from-response-in-context a precise information-theoretic characterization. In essence, choice-in-context determines a dual information source. It is further possible to place that characterization within the realm of recent developments which propose the coevolutionary mutual entrainment – in a large sense – of different information sources to create larger metalanguages containing the original as subdialects. This work also permits treating gene-culture and brain-culture condensations using a similar, unified, conceptual framework of information source coevolutionary condensation. The Atlan and Cohen version of the immune cognition model suggests, then, the possibility that human culture and the human immune system may be jointly convoluted: To neuroimmunology and immunogenetics we add immunocultural condensation.

The evolutionary anthropologists' vision of the world, as we have interpreted it, sees language, culture, gene pool, and individual CNS and immune cognition as intrinsically melded and synergistic. We propose, then, that culture, as embodied in a local cognitive sociocultural network, and individual immune cognition may become a joint entity whose observation may be 'confounded' – and even perhaps masked – by the distinct population genetics associated with assortive mating due to linguistic and cultural isolation.

8.3 Punctuated interpenetration: adapting to pathogenic hierarchy

Ademi et al. (2000) see genomic complexity as the amount of information a gene sequence stores about its environment, analogous, perhaps, to a (simplified) Kolmogorov complexity representation of that environment. Something similar can be said of a reverse process: environmental complexity is the amount of information organisms introduce into the environment as a result of their collective actions and interactions (Lewontin, 2000), perhaps suggesting the necessity of defining a 'joint' Kolmogorov complexity.

From that interactive perspective we can invoke an information theory formalism, using the Rate Distortion Theorem as applied to Ademi's mapping. The result is a description of how a structured environment, through adaptation, literally writes a (necessarily) distorted image of itself onto the genetic structure of an organism in a punctuated manner through a kind of spontaneous symmetry breaking driven by increased mutual information coupling, while itself often (but not always) being affected by changes in the organism. Wallace and Wallace (1998, 1999) use punctuated splittings and coagulations of languages-on-networks to represent, respectively, speciation and coevolution.

Once immune cognition and CNS cognition are seen as linked, it is possible to embed the dual information sources associated with those cognitive processes (Wallace, 2002) within a matrix defined by a local, but larger and encompassing, cognitive sociocultural network, using network information theory: The three information source 'languages' of the layers of cognitive process interact to create a more complicated mutual information.

Note that in evolutionary process the interaction is largely through the filter of selection acting on expressed phenotypes. Pathogenic social hierarchy, viewing it as a force for epigenetic catalysis, is a more active instrument of interaction between physiological and psychosocial languages.

Through the particular influence of the local sociocultural network on humans, culture then becomes, quite literally, part of human biology.

We propose one more iteration, using the Joint Asymptotic Equipartition or Rate Distortion Theorems, that apply to dual interacting information sources. We suppose that the tripartite mutual information representing the interpenetrative coagulation of immune, CNS, and locally social cognition, is itself subjected, via an analog to epigenetic catalysis, to influence by a larger embedding, but highly structured, process representing the power relations between groups. Most typically, these would constitute hierarchical systems of imposed economic inequality and deprivation, the historic social construct of racism, patterns of wage-slavery or, very likely, a coherent amalgam of them all. Above we gave a mathematical treatment of such multiple interacting information sources in terms of network information theory.

The result of this iteration, using rate distortion arguments, is that pathologies of social hierarchy write distorted images of themselves down the chain of human biological interpenetration onto the development and functioning of the immune system at every stage of life, as instantiated through processes of nested epigenetic catalysis, described above.

We thus propose that chronic vascular inflammation resulting in coronary heart disease is not merely the result of changes in human diet and activity in historical times (e.g., Ridker, 2002), but represents the image of literally inhuman 'racial' and socioeconomic policies, practices, history, and related mechanisms imposed upon the immune system, beginning in utero, and continuing throughout the life course: epigenetic catalysis writ large on individuals and populations.

Our interpretation is consistent with, but extends slightly, already huge and rapidly growing animal model and health disparities literatures, (Bosma et al., 1997, 1998, 2001; Schapiro et al., 1998; DeGroot et al., 2001; Gryazeva et al., 2001; Kaplan et al., 1996; Wilkinson, 1996; Karlsen and Nazroo, 2002). Kaplan et al. (1996), for example, found that female primates fed an atherogenic diet were markedly graded on risk of CHD inversely according to social status, in spite of the supposed protective effect of female hormones.

One particular consequence of our importation of punctuated equilibrium formalism – via the spontaneous symmetry breaking arguments of Chapter 3 – is that the writing of pathogenic social hierarchy onto the human organism through vascular inflammation can be punctuated, accounting in a natural manner for the staged progress of the disease, particularly in view of the sensitivity to imposed stress during developmental critical peroids.

8.4 Implications for intervention

Our analysis suggests that, under conditions of racism, wage slavery, draconian socioeconomic inequality, and outright material deprivation, an aspirin a day (or some chemical equivalent) will not keep death at bay. That is, pathogenic social hierarchy is a protean and determinedly plieotropic force, having many possible pathways for its biological expression: if not heart disease, then high blood pressure, if not high blood pressure, then cancer, if not cancer, diabetes, if not diabetes, then behavioral pathologies leading to raised rates of violence, substance abuse, or high risk sexual activity, and so on. We have explored a particular mechanism by which pathogenic social hierarchy imposes an image of itself on the human immune system through vascular inflammation. Results like those of Collins and Williams (1996), McCord and Freeman (1990), or as shown in figures 7.9 - 7.11, imply, however, the existence of multiple, competing pathways along which deprivation, inequality, and injustice operate. These not only write themselves onto molecular mechanisms of basic human biology, but become, as a result of the particular role of culture among humans, literally a part of that basic biology.

The nature of human life in community, and the special role of culture in that life, ensures that individual psychoneuroimmunology cannot be disentangled from social process, its cultural determinants, and their historic trajectory. Psychosocial stress is not some undifferentiated quantity like the pressure under water, but has a complex and coherent cultural grammar and syntax that express themselves as chronic vascular inflammation.

For marginalized populations, this is not a simple process amenable to magic bullet interventions. Substance abuse and overeating become mechanisms for self medication and the leavening of distorted leptin/cortisol cycles. Activity and exercise patterns may be constrained by social pathologies representing larger-scale written images of racism (Wallace et al., 1996).

Culture, as a kind of extended generalized language, is path-dependent: Changes are almost always based on, and consistent with, preexisting structures, i.e., the burdens of history. A cultural history of pathogenic social hierarchy, then, may continue to write itself on human immune cognition as chronic vascular inflammation, requiring large-scale and very disruptive affirmative action interventions for redress. That is, elaborate, ecosystem-based programs of Integrated Inflammation Management (IIM), much like Integrated Pest Management (IPM) in agriculture, may be required. The history of fighting outbreaks of agricultural pests with magic bullet pesticide application is rife with failure, as the organisms simply evolve chemical resistance. The writing of pathogenic social hierarchy onto human immune function over the life course seems to be a fundamental, and likely very plastic, biological mechanism equally unlikely to respond, in the long run, to magic bullet interventions. Rather, an extension of the comprehensive reforms which largely ended the scourge of infectious disease in the late 19th and early 20th centuries seems prerequisite to significant intervention against coronary heart disease and related disorders for marginalized populations within modern industrialized societies.

This analysis has obvious implications for the continued decline of CHD within the US majority population. Our own studies, partly summarized in the previous Chapter, show clearly that the public health impacts of recent massive deindustrialization and deurbanization in the US have not been confined to urbanized minority or working-class communities where they have been focused, but have become regionalized in a very precise sense so as to entrain surrounding suburban counties into both national patterns of hierarchical, and metropolitan regional patterns of spatially contagious, diffusion of emerging infection and behavioral pathology (Wallace and Wallace, 1997, 1999; D. Wallace and R. Wallace, 1998). In essence, social disintegration has diffused outward from decaying urban centers, carrying with it both disease and disorder (Wallace et al., 1997). To use a phrase first coined by Greg Pappas, concentration is not containment, and the system of American Apartheid (Massey and Denton, 1992) has quite simply been unable to limit health impacts to minority communities, a reality starkly contrary to very deeply held and emotionally compelling cultural beliefs in the US.

In precisely the same sense, it seems virtually inevitable that American Apartheid, as expressed in patterns of pathogenic hierarchy entraining all subpopulations, will similarly constitute a very real biological limit to possible declines in CHD among both white and Black subpopulations. Figure 7.9 suggests that nobody is more enmeshed in, and hence susceptible to, the pathologies of hierarchy than members of a majority whose fundamental cultural assumptions include the social reality of divisions by class and race.

There is some empirical support for this perspective as it affects chronic inflammatory disease: D. Wallace and R. Wallace (1998) examined the spatial structure of diabetes mortality incidence, a correlate of CHD, in several US metropolitan regions, contrasting the time periods 1979-85 and 1986-94 at the county level. While the overall structure of diabetes mortality was poverty-driven, the New York metropolitan region, one of the most virulently segregated in the US (Massey and Denton, 1992), showed a startling decline in the strength of the relation between diabetes mortality rate and poverty rate over the two time periods, from $R^2 = 0.44, P = 0.0003$ to $R^2 = 0.16, P = 0.03$. They conclude that the marked weakening of the relation for the New York metro region is not a sign of improvement in the lot of the poor, rather it means that high incidence is spilling over into areas with low-to-moderate poverty rates, i.e., high incidence is crossing class lines. The explanation, they infer, may lie in either or both of two hypotheses: the level of stress once associated with poverty is affecting those above the poverty line in this metro region, or the response to stress once concentrated in the population below the poverty line has been adopted by those not living in poverty. Because of the great increase in the proportion of the US population which qualifies as obese, they believe that the explanation is a combination. The stress on the blue collar and white collar classes may lead them to seek relief, with that relief partly in excess food and passive pastimes.

We find that American Apartheid and similar systems are classic double-edged swords wounding both dominant and subordinate communities, placing a very real biological limit to the possible decline of coronary heart disease across the entire social spectrum. Programs of social and cultural reform affecting marginalized populations will inevitably entrain the powerful as well, to the benefit of all.

9
Cancer: a developmental perspective

9.1 Introduction

'Racial' disparities among cancer rates and virulence, particularly of the breast and prostate, are something of a mystery. For the US, in the face of slavery and its sequelae, centuries of interbreeding has greatly leavened genetic differences between African-Americans and other population sectors, but marked contrasts in disease prevalence and progression persist. Adjustment for socioeconomic status and lifestyle, while statistically accounting for much of the variance in breast cancer, only begs the question of ultimate causality. Here we propose a more basic biological explanation that extends the theory of immune cognition to include an elaborate tumor control mechanism constituting the principal selection pressure acting on pathological mutating cells or their clones.

The interplay between them occurs in the context of an embedding, highly structured, system of culturally-specific psychosocial stressors that can act as an analog to an epigenetic catalyst. According to our model, the larger embedding system of stressors can drive the disease process via enhanced risk behaviors, accelerated mutation rate, and depressed mutation control. The dynamics are analogous to punctuated equilibrium in simple evolutionary systems, accounting for the staged nature of disease progression.

Ultimately, we find that the failure of tumor control mechanisms constitutes a kind of inverse of the developmental disorder arguments of Chapter 5. Crudely, initiation and progression of cancer are seen as a normal and inevitable consequence of multicellularity (e.g., Nunney, 1999), but the normal final phenotype of a cancer's developmental trajectory, the S_∞ in the sense of Chapter 2, as it were, is extinction or containment. Environmental farming of tumor development, via mechanisms of, or similar to, epigenetic catalysis, can create a 'developmental disorder' resulting in cancer, i.e., survival of proliferating clones of cancerous cells.

We conclude that social exposures are, for human populations, far more than incidental cofactors in cancer etiology. Rather, they are part of the basic biology of the disorder.

9.2 Context

Cardiovascular disease and cancer are major causes of mortality in the United States structured by 'race', class, and gender (Herberman, 1995). Cancers of the breast and prostate - hormonal cancers - particularly show large disparities by ethnicity and economic class in incidence among young adults, stage at presentation, and mortality rate (e.g., Parker et al., 1998).

Although certain genetic alleles predispose individuals to higher susceptibility for these cancers (e.g., Gong et al., 2002, for prostate cancer), recent changes of incidence and mortality in time and geography indicate genes alone do not explain the expressed population-level patterns. At present, African-American women under age 35 suffer an approximately two-fold higher age-specific rate of breast cancer, compared to white women, and the mortality rate is about three times higher (Shavers et al., 2003). For prostate cancer, African-American men have a 2-fold higher mortality rate, and 50 percent higher incidence rate, than their white counterparts (Sarma and Schottenfeld, 2002).

During the 1980's, when ideologies of individualism particularly influenced US scientific thinking, interest in health differentials waned. In its stead, the life style doctrine arose: people get sick because they don't take responsibility for their own health. The connection between diet and breast cancer was an often-cited example of how life style affects health. However, a large literature on determinants of risk behaviors explored the bases of 'life style' decisions and found them rooted in social and economic processes (e.g., Bursick, 1986; Fullilove and Fullilove, 1989; Kane, 1981; Wallace et al., 1996; Wallace and Fullilove, 1999).

Many of the risk behaviors associated with AIDS, drug abuse, and violence, were shown to be coping mechanisms for dealing with frustration, pain, deprivation, humiliation, and danger (e.g., Ben-Tovin et al., 2002; Wallace et al., 1996). The particular modes of coping spread, first between social networks and then within social networks by branching processes (Latkin et al., 1999). Indeed, one of the classic studies of drug use, *The Heroin Epidemics*, described all these contagious small-scale processes early on (Hunt and Chambers, 1976). Risk behaviors may explain part of the pattern in hormonal cancers.

Such behaviors, however, may not totally explain population differentials in hormonal cancer incidences and mortality rates.

We propose an approach that more fully integrates the biocultural processes that shape the development of humans, their cancers, and differentials in both their susceptibility and pathways of disease progression. We begin with Nunney's (1999) evolutionary history of cancer, as opposed to more conventional local evolutionary dynamic theories of tumorigenesis within an organism (e.g., Bertram, 2001). As described above, Nunney's analysis suggests that in larger animals, whose lifespans are proportional to about the 4/10 power of their cell count, prevention of cancer in rapidly proliferating tissues becomes more difficult in proportion to their size. Cancer control requires the development of additional mechanisms and systems to address tumorigenesis as body size increases – a synergistic effect of cell number and organism longevity.

Nunney sees this pattern as representing a real barrier to the evolution of large, long-lived animals and predicts that those that do evolve have recruited additional controls over those of smaller animals to prevent cancer.

Nunney's work implies, in particular, that different tissues may have evolved markedly different tumor control strategies. All of these, however, are likely to be energetically expensive, permeated with different complex signaling strategies, and subject to a multiplicity of reactions to signals. For modern humans, large animals whose principal selective environment is other humans, this suggests a critical role for the epigenetic signal of psychosocial stress, as mediated by a local sociocultural network, i.e., an embedding cognitive social structure linked to a cultural practice and history.

Contemporary evolutionary anthropology (e.g., Durham, 1991) emphasizes that culture, largely defining what social relations are particularly helpful or stressful, has become inextricably intertwined with human biology. Recent analysis (e.g., Forlenza and Baum, 2000) suggests that psychosocial stress is a very strong signal indeed and severally affects the stages of mutation control: immune surveillance, both DNA damage and repair, apoptosis, and may actually affect rates of somatic mutation – the 'mutator phenotype' we will explore below.

Atlan and Cohen (1998) and Cohen (2000), as we have described above, go even further, finding the immune system is itself cognitive. We have extended this characterization to show how an embedding sociocultural network, the local extended family which enmeshes every human, can interact with both an individual's central nervous and immune systems as a cognitive condensation that links social to psychoneuroimmunologic function. According to our analysis, a systematic pattern of externally-imposed stressors constitutes a 'language' that can interact with this condensation. The signal of imposed coherent stress then literally writes a distorted image of itself onto the functioning of the immune system via a a number of possible mechanisms.

Here we will examine the effect of structured external stress on tumorigenesis via a version of epigenetic catalysis. We will describe the local evolution of cancer within a tissue in terms of a punctuated interpenetration between a tumorigenic mutator mechanism and an embedding cognitive process of mutation control, including but transcending immune function. The essential point is that the normal developmental trajectory of a cancer, in the presence of an effective cognitive tumor control mechanism, is from an initial set of precancerous cells at S_0 to an extinction, repair, or containment at S_∞. The 'developmental disorder' in this case is a condition at S_∞ that is not constrained or repaired.

Punctuated biological processes are found across a broad range of temporal scales. Evolutionary punctuation is a modern extension of Darwinian evolutionary theory that accounts for the relative stability of a species' fossil record between the time it first appears and its extinction (e.g., Gould, 2002). Species appear 'suddenly' on a geologic timescale, persist relatively unchanged for a fairly long time, and then disappear suddenly, again on a geologic timescale. Evolutionary process is vastly speeded up in tumorigenesis, but we believe it to be subject to a form of punctuation that accounts for the often staged nature of the disease, another version of the spontaneous symmetry breaking of Chapter 3.

In essence, a cognitive tumor control process constitutes the Darwinian selection pressure determining the fate of the (path dependent) output of a 'natural' process generating precancerous cells. Externally-imposed structured psychosocial stress then jointly increases the 'natural' mutation rate while decreasing the effectiveness of mutation control through an additional level of punctuated interpenetration. We envision this as a single, interlinked process, and, extending Nunney's work, find that culture is indeed a fundamental part of human biology, at the level of very basic biological mechanisms.

For human populations, different forms of social exposures can act as carcinogens or co-carcinogens via a kind of epigenetic catalysis. Hormonal cancers, since they explicitly involve signaling molecules, should be especially amenable to an information dynamics formalism.

The central mystery we are addressing does not involve such detailed questions as the relationship between metastatic spread and primary tumor size or the like. We are, instead, focusing on the basic biology of population-level differences in disease expression, as a kind of inverse to normal development. However, the approach does provide an explanation of the temporally staged nature of cancer, in terms of multiple phase-change-like punctuations. Thus we extend the theory of the earlier chapters, as it applies to the interaction of mutating or precancerous cells and a set of related tumor control strategies that we infer must be cognitive. This set may include, but likely transcends, immune cognition.

The literature on mutator mechanisms is vast and growing (e.g., Magnasco and Thaler, 1996; Thaler, 1999). In sum, Thaler (1999) finds it is conceivable that the mutagenic effects associated with a cell sensing its environment and history could be as exquisitely regulated as transcription. Thus a structured environment may, in a higher iteration which Tenaillon et al. (2001) characterize as second-order selection, write itself, in a punctuated manner, onto the very internal workings of evolutionary punctuation itself, with evident implications for understanding tumorigenesis.

We are, to be explicit, speculating that internal cellular mechanisms controlling cancers may be actively cognitive. This cognitive internal process may itself become linked with the structured system of external selection pressures affecting the mutator. The resulting synergism would then affect tumorigenesis by means of a punctuated interpenetration between a process of socio-cellular cognition and adaptive clonal responses. Culturally crafted, systematic patterns of psychosocial stress affect tumorigenesis in a highly plieotropic manner, via an epigenetic catalysis involving far more than just chronic inflammation. This perspective is in some contrast to currently popular biomedical inflammation models of tumorigenesis that seem reflexively reductionist (e.g., Coussens and Werb, 2002).

Under this revised paradigm, cancer becomes a complicated developmental disease of human ecology, likely to respond at the population level only to multifactorial, multiscale strategies that include redressing patterns of past and continuing social and economic injustice.

9.3 Mutator dynamics

Suppose the basic properties of tumorigenesis at some 'time' k are characterized by some very general set of observable parameters that have been appropriately coarse-grained and that we write as $A_k \equiv \{\alpha_1^k, ..., \alpha_m^k\}$. We suppose that, over a sequence of 'times', the properties can be characterized by a path $x_n = A_0, A_1, ..., A_{n-1}$ having significant serial correlations that, in fact, permit definition of an adiabatically piecewise stationary ergodic information source associated with the paths x_n. We call that source **X**. This is a far more general requirement than it seems, as we are only asking that the paths x_n reflect regularities having identifiable rules of 'grammar' and 'syntax' in their staging.

We further suppose, in the (now) usual manner, that external selection pressure – the cognitive tumor control process – is also highly structured, and has an associated dual information source **Y** in the spirit of Section 2.2 that interacts not only with the system of interest globally, but specifically with its properties as characterized by **X**. **Y** is necessarily associated with a set of paths y_n.

We pair the two sets of paths into a joint path, $z_n \equiv (x_n, y_y)$ and invoke an inverse coupling parameter, a temperature-analog K, between the information sources and their paths. This leads, by the arguments of Chapters 3 and 4, to phase transition punctuation of the mutual information splitting criterion linking the two information sources, that we write as $I[K]$. Again, this is the mutual information between **X** and **Y**, under either the Joint Asymptotic Equipartition Theorem or under limitation by a distortion measure, through the Rate Distortion Theorem, or via a network information theory argument. The essential point is that $I[K]$ is a splitting criterion under these theorems, and thus partakes of the standard homology with free energy density.

Activation of our version of the mutator then becomes itself a punctuated event.

To reiterate, Thaler (1999) has suggested that the mutagenic effects associated with a cell sensing its environment and history could be as exquisitely regulated as transcription. Our invocation of the Rate Distortion, Joint Asymptotic Equipartition, or Network Information Theory Theorems in address of the mutator necessarily means that mutational variation comes to substantially reflect the grammar, syntax, and higher order structures of the embedding structures of psychosocial stress via a kind of epigenetic catalysis that serves to channel or program the interaction between mutation, clone growth, and tumor control over the developmental lifetime of the cancer. This involves far more than a simple 'colored noise' – stochastic excursions about a deterministic spine – and most certainly implies the need for exquisite regulation. We have thus provided a deep information theory argument for Thaler's speculation.

It is important to note, however, that mutator mechanics are only a small part of the overall picture. Certain risk behaviors are socioeconomically influenced, and, in turn, influence mutation rates and cancer initiation. Even the hormonal cancers are correlated with diet, exercise, alcohol intake, etc. Obesity is a very significant risk factor for breast cancer. Ulitmately, mutator dynamics are far more complicated than we are able to describe here, and a more complete story must involve describing linkages with embedding sociocultural phenomena.

Thaler (1999) further argues that the immune system provides an example of a biological system which ignores conceptual boundaries between development and evolution. While evolutionary phenomena are not cognitive in the sense of the immune system, they may still significantly interact with development. The very reproductive mechanisms of a cell, organism, or organization may become closely coupled with structured external selection pressure.

Thaler (1999) specifically examines the meaning of the mutator for the biology of cancer, that, like the immune system it defies, is seen as involving both development and evolution. In our version of the mechanism, the sudden phase transition-like change in the mutual information $I[K]$ at some critical coupling K_C with the external 'program' of imposed psychosocial stress, might represent an initiating event, while subsequent closely linked paths that lead to malignancy could be considered a series of promoting phase transitions, also taking place under the effect of the embedding epigenetic program of stress. In reality, there would seem to be a single, undifferentiated, interlinked process representing the staged failure of a cellular cognitive control strategy that can itself become convoluted with systems of structured external stressors affecting the mutator. We expand on this point:

Various authors have argued for non-reductionist approaches to tumorigenesis (e.g., Baverstock, 2000; Waliszewski et al., 1998), including psychosocial stressors as inherent to the process (Forlenza and Baum, 2000). What is clear is that, once a mutation has occurred, multiple systems must fail for tumorigenesis to proceed. It is well known that processes of DNA repair (e.g., Snow, 1997), programmed cell death – apoptosis – (e.g., Evan and Littlewood, 1998), and immune surveillance (e.g., Herberman, 1995; Somers and Guillou, 1994) all act to redress cell mutation. The immune system is believed to be cognitive, and equipped with an array of possible remediations. It is, then, possible to infer a larger, jointly-acting mutation control process incorporating these and other cellular, systemic, and social mechanisms. This clearly must involve comparison of developing cells with some internal model of what constitutes a 'normal' pattern, followed by a choice of response: none, repair, programmed cell death, or full-blown immune attack. The comparison with an internal picture of the world, with a subsequent choice from a response repertoire, is, from the Atlan/Cohen perspective, the essence of cognition.

We are, then, led to propose that a mutual information may be defined characterizing the interaction of a structured system of external selection pressures with the 'language' of cellular cognition effecting mutation control. Under the Joint Asymptotic Equipartition, Rate Distortion, or Network Information Theorems, that mutual information constitutes a splitting criterion for pairwise linked paths which may itself be punctuated and subject to sudden phase transitions via mechanisms like those of Chapter 3.

We speculate that structured external stress can become jointly and synergistically linked both with cell mutation and with the cognitive process that attempts to redress cell mutation, enhancing the former, degrading the latter, and significantly raising the probability of successful tumorigenesis.

Elsewhere (Wallace and Wallace, 2004; Wallace et al., 2009) we argue that the staged nature of chronic infectious diseases like malaria, HIV, and tuberculosis represents an information-dynamic punctuated version of biological interpenetration, in the sense of Lewontin (2000), between a cognitive 'immunocultural condensation' and an adaptive pathogen. Here we suggest that a larger system of socio-cellular cognition related to the detection and correction of mutation forms an embedding context of adaptation pressures for mutating clones of defective cells (e.g., Bertram, 2001). Subsequent learning plateau-analog phase transitions of evolutionary punctuation, in the sense of Wallace (2002), constitute the many stages of cancer.

9.4 Implications and speculations

We have applied an elaborate mathematical modeling strategy to the problem of disparities in occurrence and progression of certain cancers between powerful and marginal subgroups. As the ecologist E.C. Pielou has argued the utility of such models lies largely in their raising of questions for subsequent empirical study, that, in a scientific context, is the only true source of new knowledge.

The speculations emerging from our model are of some interest.

We have expressed tumorigenesis in terms of a synergistic linkage of a 'language' of structured external stress with the adaptive mutator and its opposing cognitive process, mutation control, via an epigenetic catalysis affecting a kind of inverse developmental process in which the starting configuration is a set of precancerous cells and the 'normal' final configuration is repair, containment, or extinction of that set. Developmental disorder, in this picture, represents survival and proliferation of a cancerous clone.

Raised rates of cellular mutation that quite literally reflect biocultural selection pressure through a distorted mirror do not fit a cognitive paradigm: The adaptive mutator may propose, but selection, via the spectrum of possibly tissue-specific cognitive tumor control mechanisms, disposes. However, the effect of the epigenetic catalyst of structured stress on both the mutator and on mutation control, that itself constitutes the selection pressure facing a clone of mutated cells, connects the mechanisms. Subsequent multiple evolutionary 'learning plateau' phase transition events in the spirit of Section 3.3 represent the punctuated interpenetration between mutation control and clones of mutated cells and constitute the stages of disease. Such stages arise in the context of an embedding culture that affects all aspects of the disease, via the mechanisms of epigenetic catalysis.

The epigenetic effects of structured external stress on both mutation and the selection pressure facing mutated cell clones implies that reductionist magic bullets and 'life style' approaches will be of severely limited effect for marginalized human populations in the absence of highly proactive socioeconomic, political, and related interventions. Cancer plays a multidimensional chess across interacting levels of biological and social organization. To counter cancer, we'll need to play the same. Only in the full context of such broad biological control can individual-oriented strategies contribute significant impact.

Social exposures are far more than incidental cofactors in tumorigenesis: adapting Robert Boyd's aphorism, we claim that 'Culture is as much a part of hormonal cancer as are oncogenes'.

Our speculations are consistent with, but suggest extension of, a growing body of research. Kiecolt-Glaser et al. (2002), for example, discuss how chronic inflammation related to chronic stress has been linked with a spectrum of conditions associated with aging, including cardiovascular disease, osteoporosis, arthritis, type II diabetes, certain cancers, and other conditions. Dalgleish and colleagues (Dalgleish, 1999; Dalgleish and O'Byrne, 2002; O'Byrne and Dalglish, 2001) have argued at length that chronic immune activation and inflammation are closely related to the etiology of cancer and other diseases. Balkwill and Mantovani (2001) claim that, if genetic damage is the match that lights the fire of cancer, some types of inflammation may provide fuel that feeds the flames.

Dalgleish (1999) has suggested application of non-linear mathematics to examine the role of immune response in cancer etiology, viewing different phenotypic modes of the immune system – the Th1/Th2 dichotomy – as 'attractors' for chaotic processes related to tumorigenesis, and suggests therapeutic intervention to shift from Th2 to Th1. We would view such a shift in phenotype as a phase transition, triggering a subsequent developmental trajectory that might or might not lead to an extended lifespan.

Our analysis implies a complicated and subtle biology for human cancer, one in which external cultural messages – psychosocial stressors – become convoluted with both pathogenic clone mutation and with an opposing, and possibly organ-specific, variety of tumor control strategies. In the face of such a biology, anti-inflammants (Coussens and Werb, 2002) and other magic bullet interventions appear inadequate, particularly for the hormonal cancers that seem to especially characterize the contrast in power relations between groups.

Although chronic inflammation, related certainly to structured psychosocial stress, is likely to be a contributor to the enhancement of pathological mutation and the degradation of corrective response, we do not believe it to be the only such trigger. The constant cross-talk between central nervous, hormonal, immune, and tumor control systems guarantees that the epigenetic catalyst of culturally constructed external psychosocial and other stressors will act to program individual physiology in a highly plieotropic manner, with multifactorial impact on both cell mutation and tumor control.

This suggests in particular that, while anti-inflammants may indeed be of benefit for individual cases, on the whole, population-level death rates from certain classes of cancer and the related disease guild of inflammatory chronic diseases will continue to express an image of imposed patterns of pathogenic social hierarchy and related deprivations like Figure 7.9. In particular, anti-inflammant or other magic bullet therapies will not be effective in reducing population-level health disparities.

It is clear, however, that amelioration of structured patterns of stress through legislation and public policy should be a priority if we are truly serious in addressing those disparities. Such a program would greatly benefit both powerful and marginalized groups, since cultural patterns of deprivation, discrimination, and pathogenic social hierarchy necessarily enmesh all, as Figures 7.7 and 7.8 suggest.

10
Autoimmune disorders

10.1 Introduction

The primary focus of this chapter will be on systemic lupus erythematosus (SLE), a multisystem autoimmune disorder that most frequently affects young women, although we will later make some remarks regarding asthma and its relation to life stress and synergism with air pollution.

In SLE, arthritis, skin rash, osteoporosis, cataracts, accelerated atherosclerotic vascular disease (ASVD), central nervous system dysfunction, and renal disease, are the most common manifestations. The severity of symptoms may markedly fluctuate over time, and the damage of the disease is of 'Type III', i.e., mediated by immune complexes ranging from just a few molecules to relatively huge structures involving whole cells coated or cross-linked by antibody. This fact accounts for the great variety of pathology seen in SLE (Paul, 1999; Liang et al., 2002). The disease is characterized by polyclonal B-cell activation, elevated production of pathogenic autoantibodies, impaired immune complex clearance and inflammatory responses in multiple organs. Like asthma, the pathological cascade is marked by an imbalance between depressed Th1 cell cytokines, that promote cell-mediated immunity, and enhanced Th2 cell cytokines, that support humoral immunity. There is increasingly strong evidence that the cytokine Interleukin-6 (IL-6) is central to this process. IL-6 is a B-cell differentiation factor that induces the final maturation of IL-4-preactivated B cells into immunoglobulin (Ig)-secreting plasma cells (e.g., Schotte et al., 2001; Linker-Israeli et al., 1991; Cross and Benton, 1999).

Kiecolt-Glaser et al. (2002) discuss how chronic inflammation involving IL-6 has been linked with a spectrum of conditions associated with aging, including cardiovascular disease, osteoporosis, arthritis, type II diabetes, certain cancers, and other conditions. In particular the association between cardiovascular disease and inflammation, as mediated by IL-6, is related to its central role in promoting the production of C-reactive protein (CRP), an ancient and highly conserved protein secreted by the liver in response to trauma, inflammation, and infection.

As mentioned previously, CRP is a pattern recognition molecule of the innate immune system keyed to surveillance for altered self and certain pathologies, providing early defense and activation of the humoral, adaptive, immune system. It is increasingly seen as a linkage between the two forms of immune response (Du Clos, 2000; Volanakis, 2001).

As Cross and Benton (1999) note, although IL-6 (and IL-10) have been most intensely studied for involvement in the pathogenesis of SLE, the cascade nature of cytokines means that all components of the cytokine network must, ultimately, be considered. We shall attempt to model this in a very general way below.

Within the US, SLE disproportionately affects African-American women, and accelerated ASVD occurs in subjects who are predominately premenopausal women at an age when ASVD is rare or unusual (Liang et al., 2002; Bongu et al., 2002). Between 1979 and 1998, SLE death rates increased approximately 70 percent among African-American women aged 45-64 years (MMWR, 2002).

The basic disparity in disease occurrence is considerable, approximately four times higher in African-Americans than Caucasians (Bongu et al., 2002). Among Caucasian women, total SLE mortality has remained stable since the late 1970's at about 4.6 deaths per million, with a decline in rates in younger, and a rise in older, women. Among African-American women, total SLE mortality rose 30 percent to a mean annual rate of 18.7 per million, with a constant rate in younger and a rising rate in older women. The rising disparity involves both increasing prevalence and worse disease in younger African-American women (Bongu et al., 2002).

Parks et al. (2002) find that the increased risk of SLE in African-Americans cannot be explained by hormonal or reproductive risk factors (i.e., breastfeeding, preeclamsia), occupational exposures (i.e., silica, mercury), medication allergy, herpes zoster, or similar factors. They suggest, rather, a central role for such personal and social stressors as racism and poverty in creating the disparity, a fundamental insight that we will explore at more length below.

The essence of the argument is recognition that immune cognition is not a simple physical phenomenon whose zero-order reference mode is a minimum energy state to which the system will automatically return, like a collection of weights on springs left to itself: all states of the immune system are, relatively speaking, rather active high energy states. We infer, then, the necessity of a cognitive decision by the immune system as to which of the possible basic modes of the immune response is to be taken as the zero-order-reference to which the system is reset, i.e., the normal pattern of self-recognizing maintenance activities of the immune system. This can be imposed via a kind of programming akin to the epigenetic catalysis of Chapter 4.

We will express deviations from the zero-order reference state of immune cognition in terms of a relatively few nonorthogonal eigenmodes representing complex systemic responses to applied perturbation – infection, tumorigenesis, tissue damage, and the like. These eigenmodes – autoimmune address of self-antigens – are a combination of innate and learned responses to such perturbation. The nonorthogonality implies the possibility of plieotropic excitation of several characteristic autoimmune pathologies by a single perturbation.

This line of reasoning seems analogous to Nunney's (1999) argument regarding the necessity of an elaborate tumor control strategy for large animals, since the probability of tumorigenesis grows synergistically as the 0.4 power of cell count times animal lifetime, itself dependent on animal size. Some similar power law calculation can probably be done comparing the number of possible eigenmodes of the human immune system vs. the number for mouse models of autoimmune diseases.

We have argued above that cognitive processes have dual information sources. These, through Rate Distortion Theorem (RDT) or Joint Asymptotic Equipartition Theorem (JAEPT) arguments, can become linked across levels of organization with external structured signals of one kind or another in a punctuated manner, instantiating epigenetic catalysis. Thus an appropriate external signal – an infection, chemical exposure, or pattern of psychosocial stress – can, if strong enough, via mechanisms similar to epigenetic catalysis, suddenly reset the zero-order of immune cognition to a mode different from the learned zero-order maintenance mode. That would be, then, an actively self-attacking mode. The (relatively) limited number of possible high probability activated states – nonorthogonal eigenmodes – then, accounts for the limited number of possible autoimmune diseases. Different excited eigenmodes will generally be triggered by different patterns of external signals through the catalytic mechanism.

We shall be particularly interested in the possible role of pathogenic social hierarchy as such a signal in the onset of SLE.

Onset of lupus is, in this model, followed by particular developmental pathways of disease expression modulated by the hypothalamic-pituitary-adrenal (HPA) axis, the flight-or-fight mechanism. The HPA axis, through production of adrenal glucocorticoids, can upregulate anti-inflammatory and downregulate pro-inflammatory cytokines, encouraging Th2 at the expense of Th1, immune phenotype. HPA axis activity can, then, serve to turn off inflammatory activation. We shall, as in Chapter 7, model the HPA axis as a cognitive submodule that can interact synergistically with immune cognition and embedding psychosocial stress to produce disease.

To paraphrase Sternberg (2001), both excess or inadequate stress hormone responses by the HPA are associated with disease. Excessive suppression of Th1 phenotype will enhance susceptibility to infection, while inadequate suppression will enhance susceptibility to inflammatory autoimmune and allergic disease. Thus chronic HPA axis overactivation, as occurs during stress, can affect susceptibility to or severity of infectious disease through the immunosuppressive effects of the glucocorticoids. In contrast, blunted HPA axis responses, for example the consequence of post traumatic stress disorder (PTSD), may enhance susceptibility to autoimmune disease.

We shall, again, model the HPA axis as a cognitive system involving a form of signal transduction driven by the magnitude of average applied stress, much like a stochastic resonance. Small averaged stress enhances or sensitizes system response, while large average stress is an overwhelming, meaningless noise that inhibits the system, causing a blunted HPA response. Autoimmune disease becomes, then, an interaction between reset-to-zero cognition and HPA axis cognition, as modulated by the signal of applied external stress, that writes an image of itself on both initiation and promotion of disease via a catalytic mechanism.

10.2 Nonorthogonal eigenmodes of immune cognition

Infection, chemical exposure, stress, or physical injury, will perturb the cognitive immune system. After successful address, it will return to the zero order state programmed into it by external information sources via the epigenetic catalysis of Chapter 4.

We assume that the immune response can be represented by some elaborate system of nonlinear relations, possibly involving cytokine, self-antibody, and anti-self-antibody concentrations and their spatial distributions. Our principal assumption is that the zero order state is programmed via mechanisms similar to epigenetic catalysis, and that changes about that state induced by external perturbations are relatively small.

We further assume that, expanding about the reference state, all variables, x_i depend on all others nearly linearly, so that we can write, to first order at time t, a system of empirical regression equations describing the cascade of cytokines:

$$x_i(t) = \sum_{j \neq i}^{s} b_{i,j} x_j(t) + \epsilon_i(t, x_1(t), ..., x_s(t)).$$

(10.1)

At reference, the x_i are defined to be zero, as are the ϵ_i. Most critically, we will assume the $b_{i,j}$ have been determined from empirical regression relations. This assumption provides the mathematical foundation for our analysis.

Here the $x_j, j = 1, ..., s$ are taken as both independent and dependent variables involved in the inevitable cytokine feedback cascade about the programmed reference configuration, and the ϵ_i represent error terms that are not necessarily small in this approximation. s may be fairly large, depending, presumably, on the size of the animal according to some power law. Note that the ϵ terms will become external perturbations in the subsequent analysis.

In matrix notation this set of equations becomes

10.2 Nonorthogonal eigenmodes of immune cognition

$$X(t) = \mathbf{B}X(t) + U(t)$$

(10.2)

where $X(t)$ is an s-dimensional vector, $\mathbf{B}$ is an $s \times s$ matrix of regression coefficients having a zero diagonal, and U is an s-dimensional vector containing error terms that are not necessarily small. We suggest that, on the timescale of applied perturbations and initial responses, the $\mathbf{B}$-matrix remains relatively constant.

This structure, by virtue of its determination through least squares linear regression, has a number of interesting properties that permit estimation of the effects of a perturbation. Rewriting, we obtain

$$[\mathbf{I} - \mathbf{B}]X(t) = U(t)$$

(10.3)

where $\mathbf{I}$ is the $s \times s$ identity matrix and, to reiterate, $\mathbf{B}$ has a zero diagonal. We next reexpress matters in terms of the eigenstructure of $\mathbf{B}$.

Let $\mathbf{Q}$ be the matrix of eigenvectors that diagonalizes $\mathbf{B}$ (or at least reduces it to block-diagonal Jordan canonical form). Take $\mathbf{Q}Y(t) = X(t)$ and $\mathbf{Q}W(t) = U(t)$. Let $\mathbf{J}$ be the diagonal matrix of eigenvalues of $\mathbf{B}$ so that $\mathbf{B} = \mathbf{QJQ}^{-1}$. The eigenvalues of $\mathbf{B}$ can be shown to all be real (D. Wallace and R. Wallace, 2000). Then, for the eigenvectors Y_k of $\mathbf{B}$, corresponding to eigenvalues λ_k,

$$Y_k(t) = \mathbf{J}Y_k(t) + W_k(t).$$

(10.4)

Using a term-by-term shorthand for the components of Y_k, this becomes

$$y_k(t) = \lambda_k y_k(t) + w_k(t).$$

Define the mean of a time dependent function $f(t)$ over the time interval $[0, \Delta T]$ as

$$E[f] \equiv \frac{1}{\Delta T} \int_0^{\Delta T} f(t) dt.$$

(10.5)

We assume an appropriately rational structure as $\Delta T \to \infty$; in particular we can, if necessary, take the perturbing signals to be the output of a stationary, ergodic information source, and in general require that all of the resulting signals will reflect that structure, so that time averages can generally be replaced by ensemble averages on the appropriate timescales.

The variance $V[f]$ over the same time interval is defined as $E[(f - E[f])^2]$. Taking matters again term-by-term, we obtain

$$V[(1 - \lambda_k)y_k] = V[w_k]$$

so that

$$V[y_k] = \frac{V[w_k]}{(1 - \lambda_k)^2}$$

or

$$\sigma(y_k) = \frac{\sigma(w_k)}{|1 - \lambda_k|},$$

(10.6)

where σ represents the standard deviation.

The y_k are the components of the eigentransformed immune system variates, and the w_k are the similarly transformed variates of the driving externalities of infection, injury, or stress. These signals are not, in particular, always likely to be random, and may have highly structured internal serial correlations of grammar and syntax, although this assumption is not necessary here.

The eigenvectors Y_k are characteristic but non-orthogonal combinations of the original variates X_k whose standard deviation is that of the transformed externalities amplified by the term $1/|1 - \lambda|$. Characteristic patterns of perturbation w can therefore trigger characteristic, but nonorthogonal, amplified patterns of general immune response Y_k, that must, among other things, instantiate the immune system's perception of the self. Although there may be s of these excited eigenmodes – in addition to the zero reference state – relatively few of them will be highly probable. It is these highly probable Y_k that we propose form the possible set of defined zero states of the immune system, one of which, including the initial $Y_0 = 0$ state, must be chosen by a cognitive reset-to-zero module, that is, forced by a programming mechanism akin to epigenetic catalysis to be the lowest energy state of the system.

Note in particular that the nonorthogonal nature of the eigenstructure of **B** implies plieotropy, i.e., that a single input signal may have multiple possible outputs, here in proportion to the magnitude of the excitation. Such plieotropy, in this context, suggests the possibility of comorbidity in autoimmune diseases.

Extension of the model to include rates of change of cytokine concentrations and the like, in addition to their magnitudes, is algebraically complicated but seems fairly direct. Intuitively, such extension must give a first order matrix relation much like equation (10.1), but now in terms of both the x_j and their time rates of change $\dot{x}_j$. Such matrix equations typically have eigenmodes with complex eigenvalues representing linked patterns of dynamical limit cycles, rather than simple fixed eigenstructures. Thus the problem becomes one of perturbation from a reference pattern of limit cycles, and of characterizing the behavior of a cognitive submodule permitting identification of that reference pattern.

10.3 Circadian and other cycles

If immune cognition could be entirely described by a simple system of first order differential equations, expanded near a zero-state, then under perturbation we would have something like

$$\dot{X}(t) = \mathbf{R}X(t) + \epsilon(t)$$

(10.7)

where $X(t)$ is the vector of displacements from the zero state, $\epsilon(t)$ the vector of perturbations, and **R** an appropriate fixed matrix of real numbers, here having purely imaginary eigenvalues. Thus the trace of **R**, the sum of the real parts of the eigenvalues, is zero, analogous to the condition on the B-matrix above. This is a simple version of the famous Langevin equation, with a full solution in terms of the Fokker-Planck equation if $\epsilon(t)$ has appropriately random white noise properties, an analysis which moves rapidly into the realm of stochastic differential equations. Unfortunately, $\epsilon(t)$ is unlikely to be random in the sense necessary for such an approach, that requires the covariance of ϵ between different times to be proportional to a delta function. We would, on the contrary, generally expect $\epsilon(t)$ to be the output of an appropriately regular information source, with elaborate covariance structure.

We can, however, make a simplified treatment quite like the previous development, provided the system is asymptotically bounded in the sense that, for all time t, there is a fixed, positive real number c such that, for all components of the vector X,

$$|x_j(t)| \le c.$$

(10.8)

That is, the system cannot move arbitrarily far from its zero state. Then, taking the time average of equation (10.7) in the sense of the previous development gives

$$E[\dot{X}] = \mathbf{R}E[X] + E[\epsilon],$$

(10.9)

where we very explicitly do not assume the perturbations are random with zero mean. Implicitly, then, we are assigning a grammar and syntax to the perturbing structures, and, if the associated information source is ergodic, we can replace the time averages with ensemble averages, across the probability space of the language of perturbation.

Writing out the left hand side of the equation gives, component by component

$$E[\dot{x}_j] = \lim_{\Delta T \to \infty} \frac{1}{\Delta T} \int_0^{\Delta T} [dx_j/dt] dt$$

$$= \lim_{\Delta T \to \infty} \frac{1}{\Delta T} [x_j(\Delta T) - x_j(0)].$$

This expression is bounded both above and below by

$$\lim_{\Delta T \to \infty} \frac{2c}{\Delta T} = 0,$$

and is thus itself zero.
We obtain, then,

$$\mathbf{R}E[X] = -E[\epsilon].$$

(10.10)

Somewhat heuristically, to first order an eigentransformation in terms of $\mathbf{R}$ gives a result analogous to equation (10.6), again component-by-component:

$$E[y_k] = -\frac{E[w_k]}{\omega_k}.$$

(10.11)

y_k is the k-th eigenvector of $\mathbf{R}$, w_k is the *eigentransformed* perturbation, and ω_k is the frequency of the cyclic eigenmode y_k.

The equation states that cyclic eigenmode amplification by nonrandom structured external perturbation is inversely proportional to eigenmode frequency. That is, slower cycles are amplified by perturbation more than rapid ones, in proportion to their period. Again, there is no orthogonality constraint on the eigenvectors of $\mathbf{R}$, suggesting the possibility of plieotropic response.

This is an interesting result: Many physiological cycles are characterized by several relatively slow processes, in comparison with the standard physiological clock of the heartbeat: daily circadian, monthly hormonal, and annual light/temperature cycles. Pathologically amplified (nonorthogonal) eigenmodes – displacements from zero related to autoimmune disease – according to this argument, may well be intimately associated with these cycles. The monthly hormonal cycle of non-menopausal women might then be related to a particular form of non-zero offset, i.e., an excited mode representing a particular autoimmune disease. Similarly, tropical populations could suffer less from excited modes associated with annual cycles of light and temperature (and their ecological sequelae), perhaps accounting for the tropical gradient in multiple sclerosis, an autoimmune disease of the central nervous system.

That is, autoimmune disease might well be classifiable by associated cycle or cycles, as well as by perturbation-of-onset.

Most autoimmune diseases would seem, of necessity, to be particularly related to the circadian cycle, which is universal, very powerful, and always fairly long compared to the heartbeat. Thus autoimmune diseases may, from this development, be especially stratified by their disturbance in various circadian rhythms (e.g., Lechner et al., 2000; Hilty et al., 2000).

Schubert et al. (1999) report a particularly interesting experiment involving a long time series of daily monitoring of urine neopterin in a white European patient with SLE. The particular focus was the association of concentration spikes with the grammar and syntax of life stressors. Urine neopterin above 300 μ-mol is predictive of SLE activity. Not all daily stressors, some quite extreme, caused significant spiking. For this patient, a raised neopterin level was triggered only by incidents of unusual conflict with close family members: thus only a very few characteristic stresses were 'meaningful' in the context of SLE, most others were not. This work strongly suggests the validity of an 'information' approach.

10.4 The retina of the immune response

A slight variation of the model above leads to further interesting speculations. Rather than taking a differential equation approach, we follow the suggestion of Schubert et al. (1999) and suppose that the daily circadian or some other cycle imposes a periodic temporal structure on the immune system, and we measure in units of that period in the sense that the state of the immune response at some time $t+1$, which we write X_{t+1}, is assumed to be a function of its state at time t:

$$X_{t+1} = \mathbf{R}_{t+1} X_t.$$

(10.12)

If X_t, the state at time t, is of dimension m, then $\mathbf{R}_t$, the manner in which that state changes in time (from time t to $t+1$), has m^2 components. If the state at time $t = 0$ is X_0, then iterating the relation above gives the state at time t as

$$X_t = \mathbf{R}_t \mathbf{R}_{t-1} \mathbf{R}_{t-2} ... \mathbf{R}_1 X_0.$$

10.4 The retina of the immune response

(10.13)

The state of the body is, in this picture, essentially represented by an information-theoretic path defined by the stochastic sequence in $\mathbf{R}_t$, each member having m^2 components: the grammar and syntax of how things change tells us much about how we are. That sequence is mapped onto a parallel path in the states of the immune response, the set $X_0, X_1, ..., X_t$, each having m components.

If the state of the body can, in fact, be characterized as an information source – a generalized language – so that the paths of $\mathbf{R}_t$ are autocorrelated, then the autocorrelated paths in $X(t)$ represent the output of a parallel information source that is, Rate Distortion arguments to the contrary, apparently a greatly simplified, and thus grossly distorted, picture of that body.

This may not necessarily be the case.

Let us examine a single iteration in more detail, assuming now that there is a zero reference state, $\mathbf{R}_0$, for the sequence in $\mathbf{R}_t$, and that

$$X_{t+1} = (\mathbf{R}_0 + \delta\mathbf{R}_{t+1})X_t,$$

(10.14)

where $\delta\mathbf{R}_t$ is small in some sense compared to $\mathbf{R}_0$.

We again invoke a diagonalization in terms of $\mathbf{R}_0$. Let $\mathbf{Q}$ be the matrix of eigenvectors which (Jordan) diagonalizes $\mathbf{R}_0$. Then we can write

$$\mathbf{Q}X_{t+1} = (\mathbf{Q}\mathbf{R}_0\mathbf{Q}^{-1} + \mathbf{Q}\delta\mathbf{R}_{t+1}\mathbf{Q}^{-1})\mathbf{Q}X_t.$$

If we take $\mathbf{Q}X_t$ to be an eigenvector of $\mathbf{R}_0$, say Y_k, with eigenvalue λ_k, we can rewrite this equation as a spectral expansion,

$$Y_{t+1} = (\mathbf{J} + \delta\mathbf{J}_{t+1})Y_k \equiv \lambda_k Y_k + \delta Y_{t+1} =$$

$$\lambda_k Y_k + \sum_{j=1}^{n} a_j Y_j,$$

(10.15)

where $\mathbf{J}$ is a (block) diagonal matrix, $\delta \mathbf{J}_{t+1} \equiv \mathbf{QR}_{t+1}\mathbf{Q}^{-1}$, and δY_{t+1} has been expanded in terms of a spectrum of the eigenvectors of $\mathbf{R}_0$, with

$$|a_j| \ll |\lambda_k|, |a_{j+1}| \ll |a_j|.$$

(10.16)

The essential point is that, provided $\mathbf{R}_0$ is chosen or tuned so that this condition is true, the first few terms in the spectrum of the plieotropic iteration of the eigenstate will contain most of the essential information about the perturbation. We envision this as similar to the detection of color in the retina, where three overlapping non-orthogonal eigenmodes of response suffice to characterize a vast plethora of color sensation. Here, if such a spectral analysis is possible, a very small number of eigenmodes of the immune response would suffice to permit identification of a vast range of perturbed bodily states: the rate-distortion constraints become very manageable indeed.

This is a necessarily more complex process than color detection since the immune system has both innate and learned components, and genetic programming is of limited value. The key to the problem, we believe, would lie in the proper rate-distortion tuning of the system, i.e., the choice of zero-mode, $\mathbf{R}_0$, imposed through catalysis by an embedding information source.

Caswell (2001) provides an accessible introduction to the kind of matrix population models we have invoked.

Much of this can be subsumed by an appropriate invocation of the tuning theorem variant of the Shannon coding theorem.

10.5 Circadian-hormonal cycle synergism.

While men and women share both circadian and annual cycles, women of reproductive age are likely to find any daily physiological patterns interacting with their hormonal cycle – stereotypically 28 days in length. This potentially creates a vast complication for any reset-to-zero cognitive module, and may significantly contribute to the comparatively higher rate of autoimmune disease among women than men in 'Westernized' countries. The argument is quite direct.

We take the female hormonal cycle to be made up of m circadian cycles, so that we can, over a single such cycle, write

$$X_{t+m} = \mathbf{R}_m \mathbf{R}_{m-1}...\mathbf{R}_1 X_t$$

$$= \mathbf{Q}_1 X_t,$$

(10.17)

with

$$\mathbf{Q}_1 \equiv \mathbf{R}_m...\mathbf{R}_1.$$

We are, in effect, doing matrix algebra 'modulo m'.

Without going into details of a modulo m expansion of equation (10.17) in terms of the components of the product matrix $\mathbf{Q}$, it is evident that the question of the phase of the product expansion is of considerable importance. That is, given a cyclic expansion for the $\mathbf{Q}$, where does one start? Which day is the first day? The matricies $\mathbf{R}_h$ may differ significantly among themselves, depending on daily perturbations. In particular, as Caswell (2001, p. 351) discusses and equation (10.17) implies, the eigenvectors of the $\mathbf{R}_h$ are all different. Thus the eigenstructure of the composite $\mathbf{Q}$, and hence the definition of the zero state such that $\mathbf{Q}_t = \mathbf{Q}_0 + \delta \mathbf{Q}_t$, depends on where one begins in the cycle, i.e., the phase. There are, then, m possible choices for a $\mathbf{Q}_0$.

For women, any reset-to-zero immunological module must choose not only the appropriate daily $\mathbf{R}_0$, but choose among m possible monthly $\mathbf{Q}_0$. This is a composite choice fraught with possibilities for error which a similar immune cognitive module in males would not confront. We suggest this complexity may indeed account for the relatively higher rates of certain autoimmune diseases among women.

10.6 The cognitive HPA axis

Here, for the sake of completeness, we could well reiterate the arguments of Section 7.5. A reader not familiar with that material should read the section. In sum, Atlan and Cohen (1998) argue that the essence of cognition is comparison of a perceived external signal with an internal, learned, picture of the world, and then upon that comparison, the choice of a response from a much larger repertory of possible responses. Clearly, from this perspective, the HPA axis, the flight-or-fight reflex, is cognitive: Upon recognition of a new perturbation in the surrounding environment, emotional memory and forebrain cognition evaluate and choose from several possible responses,

(1) No action needed

(2) Flight
(3) Fight
(4) Helplessness (flight or fight needed, but not possible).

Upon appropriate conditioning the HPA axis is able to accelerate the decision process, much like the immune system has a more efficient response to second pathogenic challenge once the initial infection has become encoded in immune memory. For example, hyperreactivity in the context of PTSD is well known. We thus take the HPA axis to be cognitive in the Atlan/Cohen sense, and thus associated with a dual information source having a grammar and syntax. Again, Section 7.5 provides a detailed mathematical treatment from this perspective.

Stress, among humans, is not a random sequence of perturbations, is not independent of its perception, and may depend heavily on context via mechanisms of information catalysis. Stress, then, involves a highly correlated, grammatical, syntactical process by which an embedding psychosocial environment communicates with an individual, particularly with that individual's HPA axis, in the context of social hierarchy. We view the stress experienced by an individual as an APSE information source, interacting with a similar dual information source defined by HPA axis cognition.

Again, the ergodic nature of the language of stress is essentially a generalization of the law of large numbers, so that long-time averages can be well approximated by cross-sectional expectations. Languages do not have simple autocorrelation patterns, in distinct contrast with the usual assumption of random perturbations by white noise in the standard formulation of stochastic differential equations.

We suggest, then, that learning by the HPA axis is, again, the process of tuning response to perturbation. In turn, this suggests the possibility of another retina argument, as above.

Suppose we require a signal transduction form, an inverted-U-shaped curve of HPA response, as in Section 7.5, for example the signal-to-noise ratio of a stochastic resonance, so that, as in equation (7.9), the amplification factor becomes

$$\frac{1}{1-<\lambda>} = \frac{1/|<W>|^2}{1+b\exp[1/(2|<W>|)]}.$$

(10.18)

This places particular constraints on the behavior of the learned average of the HPA axis, and gives precisely the typical HPA axis pattern of initial hypersensitization, followed by anergy or burnout with increasing average stress, a behavior that might well be characterized as pathological resilience.

Again, variants of this model permit imposition of cycles of different length, for example hormonal on top of circadian. Typically this is done by requiring a cyclic structure in a chain of matrix multiplications.

To reiterate the arguments of Section 10.5, while the eigenvalues of such a cyclic system may remain the same, its eigenvectors depend on the choice of phase, where you start in the cycle. This could represent a source of contrast in HPA axis behavior between men and women, beyond that driven by the ten-fold difference in testosterone levels. Again, see Caswell (2001) for mathematical details.

Next we explore how the information source dual to the cognitive reset-to-zero process for fixed eigenmodes, eigenpatterns of limit cycles, or tuned spectra, can become linked in a punctuated manner with the cognitive HPA axis and a structured language of external perturbation.

10.7 Phase transitions of interacting information systems

We suppose that the reset-to-zero cognitive module of the immune system, or the output of the cognitive HPA axis, can be represented by a sequence of states in time, the path $x \equiv x_0, x_1, ...$ Similarly, we assume an external signal of infection, tissue damage, chemical exposure, or 'psychosocial stress' can be similarly represented by a path $y \equiv y_0, y_1, ...$ These paths are, however, both very highly structured and, within themselves, are serially correlated and can, in fact, be represented by information sources $\mathbf{X}$ and $\mathbf{Y}$. We assume the reset-to-zero cognitive process of the immune system and the external stressors interact, so that these sequences of states are not independent, but are jointly serially correlated. We can, then, define a path of sequential pairs as $z \equiv (x_0, y_0), (x_1, y_1), ...$

The essential content of the Joint Asymptotic Equipartition Theorem is that the set of joint paths z can be partitioned into a relatively small set of high probability which is termed jointly typical, and a much larger set of vanishingly small probability. Further, according to the JAEPT, the splitting criterion between high and low probability sets of pairs is the mutual information $I(X, Y) = H(X) - H(X|Y) = H(X) + H(Y) - H(X, Y)$.

We suppose there is a coupling parameter P representing the degree of linkage between the immune system's reset cognition and the system of external signals and stressors, and set $K = 1/P$. Then we have

$$I[K] \approx \lim_{n \to \infty} \frac{\log[N(K, n)]}{n},$$

where N is the number of jointly typical sequences of length n.

As we have argued above, the essential homology between information theory and statistical mechanics lies in the similarity of this expression with the infinite volume limit of the free energy density, and implies the existence of phase transitions analogous to learning plateaus or punctuated evolutionary equilibria in the relations between the cognitive reset mechanism and the system of external perturbations.

Elaborate developments are possible. From a the more limited perspective of the Rate Distortion Theorem we can view the onset of a punctuated interaction between the cognitive reset-to-zero mechanism of the immune system and external stressors as a distorted image of the internal structure of those stressors within the immune response.

If we wish to examine the impact of context, say the information source Z, on a larger number of interacting information sources representing cognitive modules, say the set $Y_j, j = 1..s$, we obtain, via network information theory, a JAEPT splitting criterion of the form

$$I(Y_1, ..., Y_s|Z) = H(Z) + \sum_{j=1}^{s} H(Y_j|Z) - H(Y_1, ..., Y_s, Z).$$

10.8 Autoimmune disease

According to current theory, the adapted human mind functions through the action and interaction of distinct mental modules that evolved fairly rapidly to help address special problems of environmental and social selection pressure faced by our Pleistocene ancestors (e.g., Barkow et al., 1992). As is well known in computer engineering, calculation by specialized submodules – e.g., numeric processor chips – can be a far more efficient means of solving particular well-defined classes of problems than direct computation by a generalized system. We suggest, then, that immune cognition has evolved specialized submodules to speed the address of certain commonly recurring challenges. Nunney (1999) has argued that, as a power law of cell count, specialized subsystems are increasingly required to recognize and redress tumorigenesis, mechanisms ranging from molecular error-correcting codes, to programmed cell death, and finally full-blown immune attack.

Here we argue that identification of the normal state of the immune system is a highly nontrivial task requiring a separate, specialized cognitive submodule within overall immune cognition that acts via mechansims similar to epigenetic catalysis to drive the desired zero state into the lowest local free energy configuration. This is essentially because, for the vast majority of information systems, unlike a mechanical system, there are no restoring springs whose low energy state automatically identifies equilibrium. That is, active comparison must be made of the state of the immunological system with some stored internal reference picture, and a decision made about whether to reset to zero, a cognitive process. We further speculate that the complexity of such a submodule must also follow something like Nunney's power law with animal size, as the overall immune system and the immune image of the self, become increasingly complicated with rising number of cells.

Failure of that cognitive submodule, typically via another round of information catalysis, results in identification of an excited, inflammatory, state of the immune system as normal, triggering the systematic patterns of self-antibody attack constituting autoimmune disease, and that our analysis suggests may often be related to particular physiological cycles or signals that are long compared to heartbeat rate.

In sum, since such zero mode identification (ZMI) is a (presumed) cognitive submodule of overall immune cognition, it involves a cognitive process, convoluting incoming sensory with ongoing internal memory data in choosing the zero state. The dual information source defined by this cognitive process can then interact in a punctuated manner with external information sources according to the Rate Distortion arguments above. From a RDT perspective, then, those external information sources literally write a distorted image of themselves onto the ZMI in a punctuated manner: developmental onset of an autoimmune disease whose progression is moderated by the HPA axis and its interactions with structured psychosocial stress.

Different systems of structured external signals – infections, chemical exposures, systems of structured psychosocial stress – will, presumably, write different characteristic images of themselves onto the ZMI cognitive submodule, i.e., trigger different autoimmune diseases, perhaps stratified by their relation to circadian, hormonal, or annual cycles.

Zero mode identification is a more general problem for cognitive processes. For those dubious of the regression model argument above, a brief abstract reformulation may be of interest. Recall that the essential characteristic of cognition in our formalism involves a function $h(x)$ which maps a (convolutional) path $x = a_0, a_1, ..., a_n, ...$ onto a member of one of two disjoint sets, B_0 or B_1. Thus respectively, either (1) $h(x) \in B_0$, implying no action taken, or (2), $h(x) \in B_1$, and some particular response is chosen from a large repertoire of possible responses. We discussed briefly the problem of defining these two disjoint sets, and suggested that some higher order cognitive module might be needed to identify what constituted B_0, the set of normal states and to program that choice via epigenetic catalysis.

Suppose that higher order cognitive module, that we now recognize as a kind of Zero Mode Identification, interacts with an embedding language of structured psychosocial stress (or other systemic perturbation) and, instantiating a Rate Distortion image of that embedding stress under a local free energy minimization driven by information catalysis, begins to include one or more members of the set B_1 into the set B_0. Recurrent hits on that aberrant state would be experienced as episodes of pathology.

Empirical tests of this hypothesis, however, quickly lead again into real-world regression models involving the interrelations of measurable biomarkers, beliefs, behaviors, reported feelings, and so on, requiring formalism much like that used above.

The second stage of the argument – disease progression – involves interaction of the ZMI programming module with the HPA axis in the context of a chronic stress that can blunt HPA axis response.

The tool for this is an appropriate parametization of the network information theory mutual information $I(Y_1; Y_2|Y_3)$, where Y_1 represents the dual information source of the ZMI module, Y_2 that of the HPA axis, and Y_3 the information source of the embedding structured psychosocial stress, a cultural artifact affecting both Y_1 and Y_2.

Phase transition behavior of $I(Y_1; Y_2|Y_3)$ may be very complicated indeed. This suggests that the onset and staging of autoimmune disease, and their relation to the grammar and syntax of applied stressors, will be even more degenerate and plieotropic than is usual for strictly immune system phenomena. It would seem possible to apply yet another retina argument, producing a limited spectrum of highly probable symptom patterns.

Iteration of this overall approach would permit introduction of even more linked cognitive submodules into a larger model, with concomitant increase in subtlety of behavior.

Further development would introduce the generalized Onsager relation analysis of gradient effects in driving parameters affecting system behavior between phase transitions.

10.9 An application to asthma

Other autoimmune disorders would seem to fit the paradigm of epigenetic catalysis. Recent empirical work by Miller and Chen (2006) on childhood asthma and family stress at the individual level, while limited in venue, converges on this perspective. They find

> First... specific features of a stressful experience govern whether it relates to biological outcomes [i.e., has 'meaning', in the Atlan/Cohen sense]... Acute stressors occurring in isolation do not influence gene expression [for encoding glucocorticoid receptor and β_2-adrenergic receptor in children with asthma], but when they arise in the midst of a chronic stressor, the strength of the association with outcomes is markedly accentuated. Second... stressful experiences are associated with alterations at the level of gene expression...

These results are consistent with viewing chronic stress as a catalytic information source having particular grammar and syntax that can direct gene expression in the presence of acute stressors.

At the community level, as opposed to the individual scale, Clougherty et al. (2007), in a truly heroic study, found significant association between traffic-related pollution – nitrogen dioxide – and asthma diagnosis *only among children with exposure to violence (ETV)*. They conclude that a hightened susceptibility to pollution, associated with violence exposures or fear thereof, may lead to synergistic health effects of social and physical environmental conditions.

Using our language, chronic ETV, that is highly socially and culturally structured indeed, could be said to serve as a catalyst in asthma etiology at the community level.

Work by Shankardass et al. (2009) tells a similar story regarding the way in which stress increases the effect of traffic-related air pollution on childhood asthma incidence. They conclude that parental stress increases susceptibility to new onset of childhood asthma associated with traffic-related air pollution, finding that the similarity in the pattern of susceptibility to maternal smoking in utero suggests that biological pathways common to the response to combustion products may explain this susceptibility. They call for further study to evaluate the role of stress induced by characteristics of life in low socioeconomic status environments as a potential explanation for disparities in the health impact of air pollution observed in low SES populations. They find that understanding the role of air pollution in the causation of complex diseases like asthma requires consideration of how social factors may modify the effects of environmental exposures.

10.10 Images of pathogenic social hierarchy

Mathematical models of physiological and other ecosystems – like those we present here – are notorious for their unreliability, instability, and oversimplification. As it is said, 'all models are wrong, but some models are useful'. The mathematical ecologist E.C. Pielou finds models useful 'not in answering questions but in raising them. Models can be used to inspire new field investigations and these are the only source of new knowledge as opposed to new speculation'.

Given this caveat, the speculations raised by our modeling exercise are of considerable interest.

Recent theories of coronary heart disease – CHD – (e.g., Ridker 2002; Libbey et al., 2002) identify a dynamic and progressive chronic vascular inflammation as the basic pathogenic biological mechanism, a process in which the cytokine IL-6 and C-reactive protein (CRP) play central roles. We have reviewed something of the 'IL-6' hypothesis regarding the etiology SLE. An earlier chapter identified social structures of pathogenic social hierarchy (PSH) in the US as critical in determining population-level patterns of CHD among African-American males. Historical cultural patterns of racism and discrimination were viewed as directly writing themselves onto the language of immune cognition in a punctuated Rate Distortion manner via a kind of catalysis to produce chronic vascular inflammation among subordinate populations.

Female hormones are known to be generally protective against CHD. Where, then, does the stress of PSH express itself in women? Figure 7.9 – again – is taken from material on health and hierarchy in Singh-Manoux et al. (2003). It displays, for men and women separately, self-reported health as a function of self-reported status rank, where 1 is high and 10 low rank, among some 7,000 male and 3,400 female London-based office staff, aged 35-55 working in 20 Civil Service departments in the late 1990's. Self-reported health is a highly significant predictor of future morbidity and mortality.

Remarkably, the results for men and women are virtually indistinguishable in what is clearly a kind of toxicological dose-response curve, displaying physiological response against a dosage of hierarchy which may include measures of both stress and real availability of resources (Link and Phelan, 2000).

We propose that PSH can also write an image of itself onto both a cognitive physiological module of the immune system, what we have called zero mode identification that defines the inactive state of the immune system, and onto the cognitive HPA axis, determining both onset and progression of autoimmune disease. The (relatively) protective role of female hormones against CHD, given the indistinguishability of men and women in figure 7.9, implies existence of a plastic, pleiotropic, response of the immune system and HPA axis to PSH. In essence, one has a sex-based choice of death by CHD induced by chronic vascular inflammation for African-American men, or a particular induced autoimmune disease for African-American women, SLE. A roughly similar story can probably be told regarding the increased rate of aggressively fatal breast cancer, diabetes, and other disorders in African-American women.

That is, the message of PSH in the US is written onto the bodies of African-American men and women as, respectively, elevated rates of coronary heart disease, systemic lupus erythematosus, and allied disorders of chronic inflammation.

The rise in SLE among African-American women appears to parallel the rise of asthma among US minority urban children, an increase of 50 percent since 1980 (CDC, 1996; NCHS, 1996). The geography of asthma in places like New York City closely matches the geography of public policy-driven urban burnout and contagious urban decay (Wallace and Wallace, 1998).

Although certainly a result of gene-environment-developmental interaction – the inevitable triple helix in the sense of Lewontin (2000) – rising rates and seriousness of SLE (and diabetes, asthma, and obesity) among African-American women over the last thirty years obviate simple genetic explanations: the genetic structure of the US population has not changed suddenly. What has changed, however, is the environmental and developmental context for African-Americans. In particular, since the end of World War II, processes of urban renewal, and after about 1970, of policy-driven contagious urban decay, together amounting to massive urban desertification, have left vast tracts of what had been thriving African-American urban neighborhoods looking like Dresden after the firebombing. African-Americans born since the middle 1970's have not had the stable childhood developmental environment of their predecessors. African-Americans born before the 1970's have not had a stable aging environment, affecting other developmental processes.

Again, as argued above, the contagious hollowing out of African-American neighborhoods in many major US cities – New York, Newark, Philadelphia, Cleveland, Toledo, Detroit, St. Louis, and so on – has been compounded by, and intertwined with, a massive deindustrialization driven by the aftermath of the Cold War that Ullmann (1988), Melman (1971) and others claim is a consequence of the massive diversion of scientific and engineering resources from civilian to military enterprise during the Cold War. They conclude that a healthy US manufacturing economy required at least a 3 percent annual improvement in productivity to maintain itself against foreign competition, a rate of growth fueled by our considerable technical resources through about 1965. Thereafter, increasing consumption of scientific and engineering personnel by the military and aerospace enterprises of the Cold War, and the consequent shift in technological emphasis away from the needs of the civilian economy, made this rate of improvement impossible, and rapid US deindustrialization ensued. Wallace et al. (1997, 1999) and Wallace and Wallace (1997) examine the consequences of this collapse for the US AIDS epidemic and other public health problems.

Deindustrialization, like policy-driven contagious urban decay and urban renewal that together have hollowed out most African-American urban neighborhoods, represents the permanent dispersal of social, economic, and political capital from the worst affected regions, particularly as deindustrialization is now convoluted with urban decay and, more recently, with the foreclosure crisis.

Massey (1990) explores the particularly acute effect of this phenomenon on the African-American minority, finding that the decline of manufacturing, the suburbanization of blue-collar employment, and the rise of the service sector eliminated many well-paying jobs for unskilled minorities and reduced the pool of marriageable men, thereby undermining the strength of the family, increasing the rate of poverty, and isolating many inner-city residents from accessible, middle-class occupations.

It is the massive social disintegration of the last thirty years – particularly the widespread destruction of urban minority communities – that we believe is literally writing an image of itself upon African-American women as rising rates of SLE, among other things.

As many have argued, and figures 7.7 and 7.8 suggest, health disparities are inevitably only the tip of an iceberg enmeshing powerful or majority populations into dynamics affecting the marginalized. Relative raised rates of autoimmune disease among African-American women are a red flag: Pathogenic social hierarchy may place a severe biological limit on the ultimate effectiveness of traditional medical behavioral and drug approaches to immune-related disease across all US subgroups, not merely for African-Americans.

This suggests in particular that magic bullet medical interventions against lupus or other autoimmune disorders, to be effective at the population level, must be integrated as part of a larger ecosystem strategy addressing the more basic problems of pathogenic social hierarchy and gender discrimination in the

US. African-American women are, however, doubly burdened through the synergism of historical patterns of racism with a traditional gender discrimination which may, in fact, reflect that racism within African-American communities, and should be among the first to benefit from such reforms.

The remarkable rise of both lupus and asthma in US minority communities after 1980 seems to indicate, from this perspective, the tightening of discrimination rather than any efforts at reform. Our own studies (D. Wallace and R. Wallace, 1998; R. Wallace and D. Wallace, 1997) suggest that the inevitable failure of American Apartheid to effectively shield the powerful from the forces and impacts of marginalization means that the dominant population is being brought into a dynamic of increasing pathology as well. As we have said before, nobody is more entrained into systems like figure 7.9 than the white majority in the US, that holds itself within the same structure it holds others, and would thus benefit by reform. Such, indeed, was the message of the Rev. Dr. Martin Luther King Jr., a message that appears to have a very basic biological reality.

This, and the previous three chapters, have examined what might be called group-aggregate phenomena of epigenetic catalysis in which the characteristics of a population are simply the average over it. Our data have been taken from large scale national samples that, almost inevitably, smear out finer scale phenomena. As the ecologists note, however, pattern exists at all scales of space, time, and population, and a central problem is the understanding of linkages across scales.

The next two chapters examine processes of epigenetic catalysis involving psychosocial stressors operating at smaller scales of analysis, where 'local effects' are not averaged out, but become dominant driving mechanisms. It is as if one were fishing with nets of different mesh size, and then trying to understand the qualitatively different catch communities. Eventually one must relate the structure of such samplings across mesh size, as it were.

While the highly averaged data of this and the three previous chapters appear consonant with our formal analysis, the next two chapters will suggest the necessity of significant extensions of the theory. Gene expression may well involve emergent, collective patterns associated with crosstalk between individuals, between individuals and their embedding sociocultural networks, or between the communities in which individuals are embedded. This may be analogous to the mesoscale resonance of Wallace and Wallace (2008) in which ecological keystone dynamics entrain processes at lesser and greater scales, in the sense of Holling (1992).

11
Demoralization and obesity in Upper Manhattan

11.1 Introduction

Here, following D. Wallace et al. (2003), we apply the analysis of the previous chapters to a neighborhood geographic level that is intermediate between individual and simple aggregate scales, and where multiple factors may synergistically interact in a kind of ecosystem 'mesoscale resonance' analogous to that described in Wallace and Wallace (2008) and Holling (1992).

Length and quality of life may depend on community socioeconomic structure and neighborhood conditions (e.g., McCord and Freeman, 1990). Numerous papers have appeared recently which document relationships between the health of populations or of groups of individuals and the conditions in their neighborhoods (examples: O'Campo et al, 1997; Geronimus, Bound and Waidmann, 1999; Shumow, Vendell, and Posner, 1998). Likewise, analyses of the geographic patterns of such public health problems as low-weight births, asthma, and diabetes have shown that these problems are concentrated in neighborhoods with particular socioeconomic conditions (examples: Nyirenda and Seckl, 1998; Litonjua, Carey, Weiss and Gold, 1999).

The Center for Children's Environmental Health at the Joseph Mailman School of Public Health (Columbia University), at the time of this work, was to examine the possible three-way relationship within a birth cohort between the independent variables of (1) environmental chemical exposures of the mothers and babies and (2)socioeconomic conditions of the health areas and (3) dependent variables of health outcomes (developmental deficits, asthma, and carcinogen biomarkers). Because many health outcomes have been linked to the uterine environment and to the mother's morale and stress (examples: Phillips, 1998; Svanes et al., 1998; Collins et al., 1998b), the study subjects were nested in their health areas by examining relationships between the mother's demoralization, her pre-pregnancy body mass index, household economics, and Health Area socioeconomic conditions. Again, Health Areas were designated by the New York City Department of Health decades ago as aggregates of Census Tracts with populations of about 20,000 each.

11.2 Methods

Recruitment of the Mothers

Women attending the pre-natal clinics of Columbia-Presbyterian Medical Center and Harlem Hospital were invited to participate in the program by female research workers, many of whom were bi-lingual (English and Spanish). Target populations were Dominican (either immigrant or American born) and African-American (American-born). The women had to have lived in the study zone for at least a year and be over 18 years of age and drug-free (see figure 11.1 for the primary study zone). The recruits' informed consents were ethically obtained, and the protocol of the entire program was approved by the Columbia-Presbyterian Medical Center's Independent Review Board. Recruitment occurred between 1997 and 2000.

The recruitment of study participants was remarkable in that the women chosen were restricted to a very narrow range of characteristics, given the nature of their neighborhoods. To reiterate,

[1] These were mothers who sought and received formal prenatal treatment over a considerable period of time at major New York City health care facilities, with all that implies for class and cultural conflicts.

[2] No cigarette smokers or users of illegal drugs were included, while the communities from which they were recruited had high levels of both behaviors.

[3] Although of African-American and Spanish-speaking Dominican ethnicity, the study participants felt comfortable working with White university researchers and their staff, who repeatedly asked the most intrusive of questions for a long time.

In a sense these women were very highly socialized, nearly uniform 'best case' test subjects whose variations, on average, more reflected the influences of the neighborhoods in which they were embedded than any inherent group structure, since that structure had been artificially 'flattened' by the draconian recruitment strategy. While this may well have been a deadly flaw for the asthma study that was the original purpose of the larger work, the relative uniformity of the subject population made our analysis here possible and indeed, highly fruitful.

Prenatal Questionnaire

Each recruit was administered a detailed questionnaire before the birth of the baby. The questionnaire included sections on demographics, household conditions and economics, employment, present health and health history, and the Dohrenwend demoralization probe (all 27 questions) (Gallagher et al., 1995).

Socioeconomic Data

The source of all health area level socioeconomic and health outcome data was Infoshare, a data base and geographic information system program developed by Leonard Rodberg, director of Urban Studies at Queens College. The version used by the Center contained Health Department health outcome annual data through 1996 and demographic and socioeconomic data from the

11.2 Methods 145

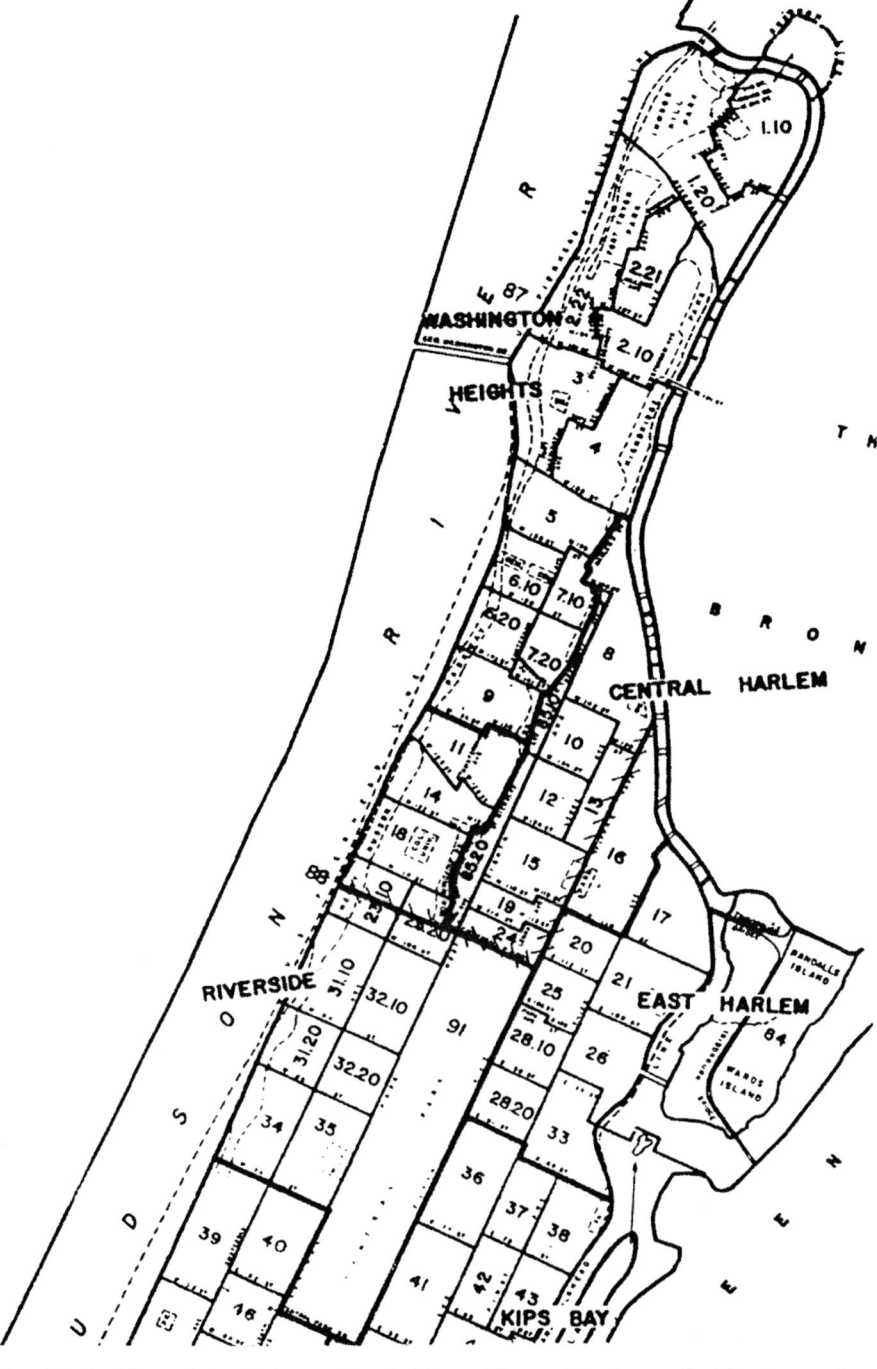

Fig. 11.1. Map of the Primary Study Zone (Upper Manhattan). The heavy lines with the light cross-lines mark the boundaries of the zone (Fifth Avenue on the East and 110 Street on the South). The numbers within the health areas are their numerical designations assigned by the New York City Department of Health and used in this study to identify the areas.

1980 and 1990 Censuses such as total population; population by age, race, and sex; median household income; total employment and employment by economic sector; and housing overcrowding. Infoshare also includes annual socioeconomic data collected by the municipal government such as number of people with public assistance.

Chronic Community Stress Index

One potential source of health disparity is chronic community level stress, stresses that do not pose a threat to life or limb but grind the residents economically, socially, or politically. An index of total chronic community stress was developed as follows:

Incidence of selected socioeconomic and social-indicator health outcome factors were calculated for 1994-1996 by summing the case numbers for the three years and dividing by the area 1990 population. Because low-weight birth is well documented in the literature as associated with maternal stress (example: Texeira et al., 1999), health area incidence of low-weight birth (1994-1996 number of births below 2500 g/10000 live births) was regressed against health area incidence of each chronic stressor. Those stressors significantly associated with low-weight birth incidence with an R^2 at or above 0.2 were included in the combined index. Thus, we included only stressors imposing a population level stress of potential importance. See Table 1 for a complete listing of the stressors by Health Area.

Table 1. Unweighted Standardized Chronic Stressers and Index of Chronic Community Stress

health area	drug*	cirrhosis*	unemploy**	%welfare	%poverty	overcrowded	1/median income	pop change	% foreign	ICCS
1.1	0.519	0.839	0.646	0.743	0.849	1.59	0.77	-0.607	1.725	53.5
1.2	0.197	0.883	0.831	1.000	0.887	0.83	0.87	-0.297	1.87	57.0
2.1	0.625	1	1.041	1.01	1.064	0.73	1	-1.372	2.41	-3.4
2.21	0.129	0.646	0.96	0.992	0.968	1.58	0.85	-1.472	2.367	-37.2
2.22	0.148	0.195	0.689	0.378	0.453	0.53	0.61	0.595	1.11	77.7
3	0.102	1.006	0.804	0.809	0.884	0.54	0.74	-1.114	2.013	-27.6
4	0.300	0.567	1.173	1.114	1.158	1.21	1.04	-1.274	2.437	-1.4
5	0.564	1.012	1.044	1.048	1	0.51	0.97	-0.765	2.152	41.8
6.1	1.029	0.687	1.017	0.978	0.981	0.48	0.89	0.39	1.44	165.0
6.2	1.055	0.231	0.862	0.995	0.92	0.59	0.99	0.328	1.841	131.9
7.1	1.320	0.889	0.847	0.942	0.977	0.88	0.99	0.648	0.823	242.9
7.2	1.253	2.227	0.716	0.84	1.055	1	1.02	1.095	0.828	297.6
8	1.286	1.095	1.109	1.212	1.146	1.17	1.12	2.484	0.297	433.8
9	0.143	0.571	1.106	1.006	1.185	1	1.01	-1.106	1.97	18.1
10	1.430	1.377	1.559	1.302	1.412	1.35	2	2.476	0.389	514.9
11	0.222	1.665	1.008	1.415	1.389	1.42	1.47	2.145	1	367.1
12	2.235	0.291	1.07	1.381	1.241	1.28	1.41	2.991	0.278	556.5
13	2.129	4.866	1.07	1.108	0.942	1.34	1.09	3.756	0.306	628.0
14	0.626	1.228	0.812	0.652	0.846	1.08	0.78	1.893	0.721	279.7
15	2.461	1.586	1.108	1.19	1.248	0.81	1.35	2.864	0.712	511.3
18	0.085	0.414	0.392	0.424	0.392	0.17	0.66	-3.198	1.351	-195.4
19	2.263	3.942	1.103	1.313	1.637	1.23	2.04	3.94	0.434	700.6
23.1	1.201	0.325	0.571	0.195	0.408	0.62	0.56	1	0.939	156.6
23.2	0.378	2.477	0.827	0.836	0.952	0.8	0.76	1.577	1.262	247.4
24	1.560	2.458	1.119	1.135	1.328	1.47	1.54	1.581	0.48	442.4
85.1	1.000	0.706	0.841	1.187	1.402	0.94	1.49	4.069	0.53	525.7
85.2	1.132	1.941	1	0.976	1.17	1.22	1.19	2.867	0.734	434.7
weighting factor	50.09	21.36	26.02	40.81	37.99	19.63	50	65.36	-57.86	

*deaths per 100,000
**number unemployed/population 17-64 years old
#(standardized%extremely overcrowded units-standardized%foreign born)/median
ICCS=Index of chronic community stress

These included:

Drug and Cirrhosis deaths per 100,000, unemployment rate, percent population on welfare, percent living in poverty, percent badly overcrowded housing, inverse median income, population change 1980-90, and percent foreign born.

Each stressor incidence of each health area was standardized by dividing it by the median incidence of the 27 health areas of the Upper Manhattan study zone. Each standardized incidence was weighted by multiplying it by the R^2 of its association with low-weight birth incidence. For each health area, the sum of all the standardized weighted incidences became the Index of Chronic Community Stress (ICCS). To check the validity of low-weight birth incidence as an indicator of stress, we also created an ICCS based on diabetes mortality incidence because type II diabetes (the vastly dominate type) has been correlated with community stress in the literature (Rajaram and Vinson, 1998). Regression of the two indices of chronic stress yielded an R^2 of 0.99: Our low birthweight-based index appears valid.

11.3 Data Analysis

All data manipulations and analyses were performed with the statistical analysis program Statgraphics Plus. The 27 separate demoralization item scores of each mother were summed to form a score of overall demoralization. The health areas were assigned to quintiles according to their ICCS's. The distributions of total individual demoralization score values within the quintiles were, with one exception, skewed normal. Average and median total scores were calculated for each quintile and plotted against the quintile weighted average ICCS of the health areas that actually had mothers living in them. Weighting was by number of mothers in the health areas. The ICCS quintiles of health areas were grouped with six health areas in the best and worst quintiles and five in each of the three middle quintiles.

Body mass index (BMI) was calculated in the standard manner by converting the height and *pre-pregnancy* weight of each recruit into kilograms and meters and applying the standard formula: weight/height*height. The standard cutoffs were also used: BMI's over 25 were considered overweight and those over 30, obese. Average and median quintile BMI's were plotted against the weighted average quintile ICCS.

The questionnaire included five questions on household deprivation: experience of food insecurity, experience of housing insecurity, cutoff of utilities due to non-payment, lack of needed clothing, and inability to pay for needed medical care/medicine. Demoralization score and BMI were separately regressed at the individual level against both individual household deprivations such as experience of food insecurity and against total number of household deprivations. Two household deprivation scores were developed for each quintile.

The first (DEPSCORE) was calculated as follows:

(percent of women with one deprivation)+(2*percent of women with 2 deprivations)+(3*percent with 3)+(4*percent with 4)+(5*percent with 5).

The second (DEPSCORE2) omitted the percent with one deprivation. Each deprivation score indicates both incidence and intensity of household economic shortfalls within the quintiles. DEPSCORE2 omits the single instances of shortfall, in case some were mere flukes, and measures only multiple deprivations.

The four following factors were, round-robin fashion, plotted against each other: weighted average quintile ICCS, quintile household deprivation scores, quintile average demoralization score, and quintile average BMI.

11.4 Modeling demoralization and chronic stress

Signal transduction is a very general pattern seen in phenomena as diverse as neural response to stimulus (e.g., McClintock and Luchinsky, 1999), physiological response to arousal (e.g., Wilken et al., 2000; Wilson et al., 2000) and in dose-response to hormonal mimetics (Bigsby et al., 1999). Recently signal transduction has come to be modeled as a 'stochastic resonance' in which a signal too weak to trigger some strong threshold response is augmented by an additive noise. If the noise is too weak, no triggering occurs. If the noise is too strong, the signal is entirely washed out by the noise. Thus the signal-to-noise ratio (SNR) of the applied signal and the much larger threshold response – effectively an amplification of the weak signal – will undergo a rise to a peak with applied noise, and then a subsequent decline as the noise overcomes the signal, according to theory, hence the characterization as a 'resonance.'

The general form for the SNR of a simple stochastic resonance, as a function of the noise amplitude x, is typically given, after some fairly arduous calculation, by the expression (Gammaitoni et al., 1998, eq. 4.51; Braiman et al., 1995; Kadtke and Bulsara, 1997; Henegan et al., 1996; McClintock and Luchinsky, 1999),

$$SNR(x) \approx \frac{1}{x^2}[1 + \alpha \exp(1/(2x))]^{-1}.$$

(11.1)

α is a scaling parameter related to the properties of the weak applied signal.

In order to model the relation between demoralization and chronic stress we used the nonlinear regression procedure of the Statgraphics statistical package to, simultaneously, linearly rescale the chronic stress index and choose the

α which best fits the demoralization data. In essence we 'argue by abduction', to use Hodgson's terminology (Hodgson, 1993), that a stochastic resonance SNR functional form will provide a good fit to 'social' as well as neural or chemical signal transduction.

11.5 Results

Demoralization and Chronic Stress

The summed demoralization scores ranged from one to 89. The 225 scores were skewed-normally distributed with an average of 34. The best health area with respect to ICCS had an index of -195; the worst health area, 701 (See Table 1 for details).

Four of the five sets of demoralization scores were skewed-normally distributed, and one, the middle quintile, multimodally. Thus, use of the average or median to represent the central value for these quintiles is reasonable. Figure 11.2, the plot of the average summed demoralization scores of the quintiles against the weighted average chronic stress index, shows an inverted, asymmetric 'U' shape. Mann-Whitney nonparametric comparison of average rank of the summed demoralization scores of the middle with the bottom quintile and of the top with the bottom quintile yielded a statistically significant difference ($P = 0.02$) for the former and no difference for the latter ($P = 0.25$). A two-tailed t-test also showed a difference between the demoralization scores of the middle and bottom quintiles ($P = 0.01$). The asymmetry of the inverted 'U' resulted in no statistical difference between the top and middle quintiles. However, from the differences detected, the shape of this curve appears to be real and not a result of chance. The basic descriptive statistics such as mean, median, variance, and upper and lower quartiles also indicate differences between these quintiles. See Table 2 for details.

Since the two worst chronic stress quintiles are composed of health areas with very high proportions of African-American population, the potential difference in demoralization scores between Dominican and African-American recruits had to be explored. Five health areas were home to both Dominican and African-American recruits. T-test, Mann-Whitney nonparametric test, and Kolmogorov-Smirnov nonparametric test showed no statistically significant difference between the demoralization scores of the Dominican and African-American women living in these five health areas. No significant difference was found between the demoralization scores of native-born and immigrant recruits living in the same nine health areas.

In most large cities of the USA, housing consumes a large percent of the household income, larger than the recommended 25%. Regression of the health area ICCS against the 1990 median rent reveals a close negative relationship with an R^2 of around 0.7. The rents were, thus, partly determined by the community conditions. The two worst quintiles had the lowest weighted av-

Table 2. Descriptive Statistics of Demoralization Scores and BMIs by Quintile

Demoralization Scores

Quintile	1	2	3	4	5
Average	34.78	34.8	39.95	32.23	28.51
median	31	33	42	31	26
geometric mean	29.86	31.09	32.32	30.33	25.59
standard deviation	16.9	16	19.72	16.06	13.42
minimum	1	7	2	6	8
maximum	80	89	81	77	69
lower quartile	21.5	23	25	25.5	18
upper quartile	46	42	55	43.5	35

Body Mass Indices

quintile	1	2	3	4	5
average	24.6	24.78	26.96	26.8	26.82
median	23.46	22.86	25.42	25.75	25.66
geometric mean	24.02	24.06	26.74	26.03	26.01
standard deviation	5.9	6.48	3.57	6.59	6.86
minimum	15.69	17.01	20.97	16	17.73
maximum	55.03	45.59	33.01	45.31	42.09
lower quartile	20.71	19.53	24.27	22.01	21.28
upper quartile	27.49	28.42	30.62	30.68	32.88

erage median rents, and the residents of the two best quintiles paid over 100 dollars a month more than those of the two worst quintiles in 1990.

Signal Transduction Modeling

Figure 11.3 shows the result of the nonlinear regression fitting of the relation between quintile demoralization and ICCS to the stochastic resonance SNR function.

Some 73 % of the variance, as adjusted for the degrees of freedom, is accounted for by this model. The asymmetric stochastic resonance SNR relation is a very good fit indeed for our data.

Body Mass Index

The BMI was skewed normally distributed for the over 200 women with complete height and weight data with an average of 25.68, a minimum of 15 and a maximum of 55. The distributions for each quintile, however, were not normally distributed and necessitated use of nonparametric tests for comparison. Figure 11.2 displays both the quintile average demoralization scores and the quintile average BMI plotted against the weighted quintile average ICCS. The BMI appears to show a step function at the middle quintile. The BMI's of the middle quintile are significantly different from those of the best quintile in both the Mann-Whitney test ($P = 0.009$) and the Kolmogorov-Smirnov test ($P = 0.005$). Table 2 displays the basic descriptive statistics of the BMI's of the quintiles.

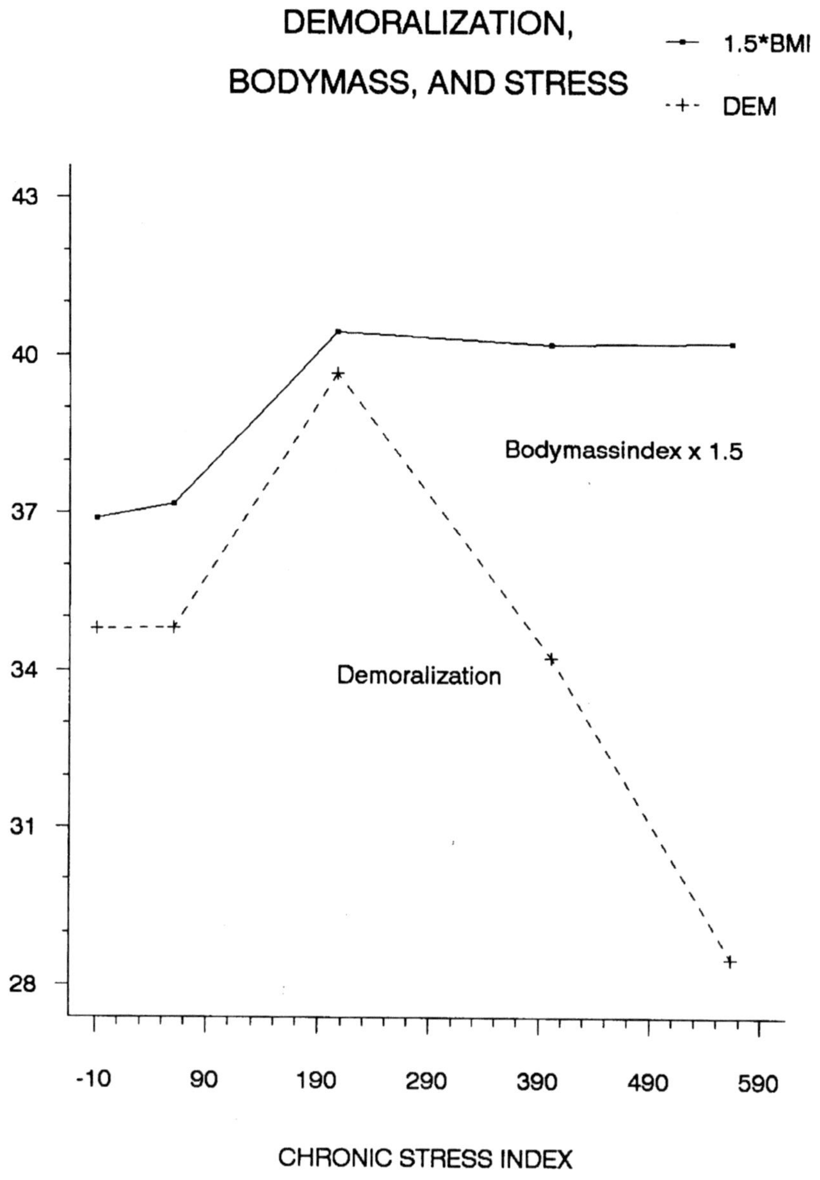

DEMORALIZATION, BODYMASS, AND STRESS

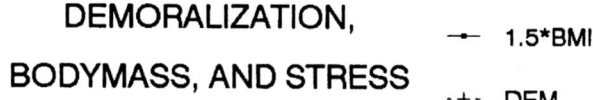

Fig. 11.2. Average Demoralization Scores and Average BMI's of the Chronic Community Stress Quintiles vs the Quintile Average Chronic Stress Indices. Note the inverted 'U' shape with peak in the middle for the average demoralization scores. Note the step function of the average BMI's with the high plateau of quintiles 3-5.

11 Demoralization and obesity in Upper Manhattan

Table 3. Summary of Data for the Chronic Community Stress Quintiles

quintile	weighted ave. ICCS	DEPSCORE	DEPSCORE2	wted ave median rent 1990 ($)	% with food insecurity	% with housing insecurity	ave. people per household	ave. demoral total	ave. BMI
1	-9.3	93.4	77.6	438.4	22.37	23.68	2.99	34.77	24.6
2	60.8	117.4	89.1	425.2	15.22	28.26	2.7	34.8	24.78
3	208.1	85	65	376.8	20	20	3	39.65	26.96
4	400.9	49.9	36.3	344.7	11.36	9.09	3.25	34.23	26.8
5	563.4	56.5	20.6	314.8	5.13	10.26	3.21	28.51	26.82

There was no statistically significant difference in the BMI's of Dominican and African-American women living in the same health areas. Indeed, both Dominican and African-American women live in the health areas of the middle quintile.

Household Deprivation

The majority of recruits reported none of the five household deprivations (126 out of the 225). The others reported numbers of deprivations as follows:

```
One deprivation  : 49 recruits
Two deprivations : 24 recruits
Three            : 16 recruits
Four             :  9 recruits
Five             :  1 recruit.
```

At the individual level, the number of household deprivations was significantly associated with total demoralization score ($P < 0.00001$) but with an R^2 of only 0.15. This was the only individual level association with an $R^2 > 0.1$. The number of household deprivations of the individual recruits was negatively associated with the ICCS of the health areas ($P = 0.0075$) but the R^2 was very small (0.032). Number of household deprivations had no association with BMI ($R^2 = 0$).

For quintile deprivation scores, see Table 3. The household deprivation scores of the two worst quintiles with respect to ICCS were much better than those of the two best quintiles. Thus, household conditions contrasted with community conditions.

Household economics also depends on the number of people sharing the resources and contributing to the resources. The numbers in the individual economic household ranged from 1 to 13. At quintile level, the average number of people per economic household was strongly negatively associated with the two deprivation scores: DEPSCORE ($P = 0.002, R^2 = 0.97$) and DEPSCORE2 ($P = 0.033, R^2 = 0.824$). At the quintile level, BMI showed a positive trend with average number of people in the economic household: $P = 0.17, R^2 = 0.52$.

In examining the five different forms of deprivation, strong associations were found between all but one and demoralization score. The one not associ-

11.5 Results

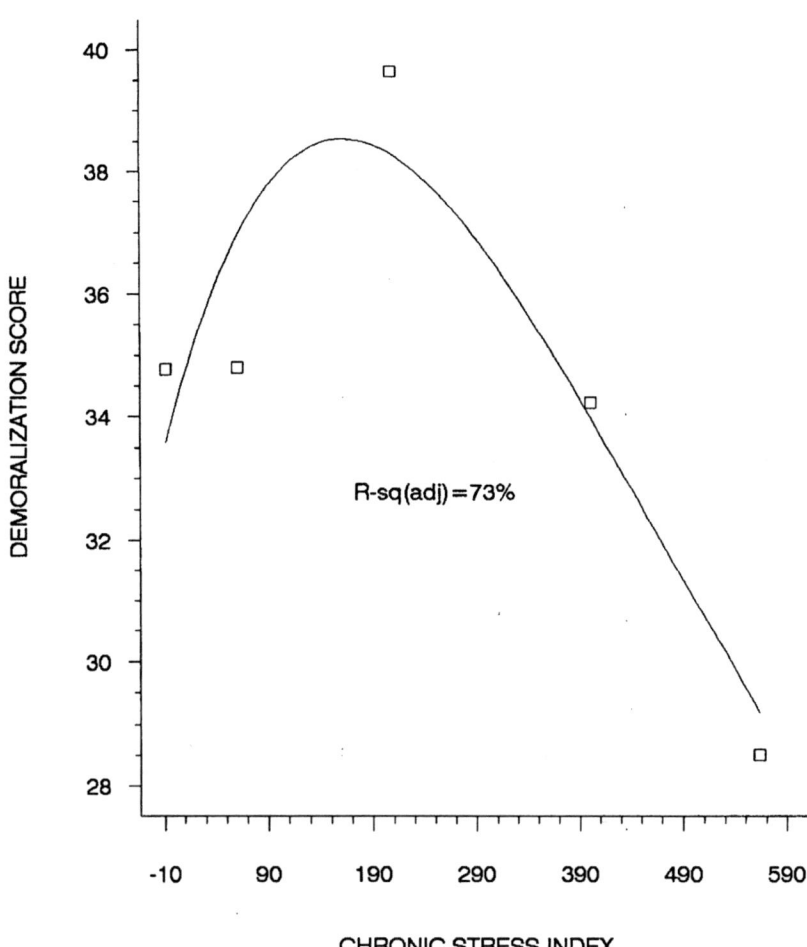

Fig. 11.3. Fit of the stochastic resonance signal-to-noise ratio/signal transduction model to the quintile average demoralization scores and quintile weighted average community chronic stress indices, based on 225 recruits. The data are fit by nonlinear regression to equation 11.1.

ated with significantly higher demoralization scores (t-test) was utility cut-off. Food insecurity and housing insecurity were the two strongest associations:

Deprivation	Average demoralization score	
	With the deprivation	without
Food insecurity	47.7	31.5
Housing insecurity	46.0	31.2.

Table 3 displays by quintile the percent of recruits reporting food insecurity, housing insecurity, average economic household size, average demoralization score, and average BMI. The second best quintile showed a decidedly smaller average household size and by far the highest proportion of recruits with housing insecurity.

11.6 Discussion

The Asymmetric Inverted 'U' Curve

It is not reasonable to interpret the curves formed by plotting stressor quintile averages against the quintile average demoralization scores at their face values. As stress increases, women do not get better morale. The inverted 'U' shape is reminiscent of the dose/response curves of hormonally active pollutants (Andersen et al., 1999, for example) that act by either mimicking or antagonizing naturally occurring hormones. For a detailed explanation of the inverted 'U' curve of endocrine disrupter dose/response, see Bigsby et al., 1999. We have two possible explanations for the inverted U shape which are not mutually exclusive.

Explanation 1: Like endocrine disrupting pollutants, the combined and individual chronic community stressors may act as a signal-boosting 'noise'. Above a certain level of chronic stress, the mothers or their social networks stopped responding actively to community conditions and, overwhelmed by the 'noise', withdrew from community involvement.

Explanation 2: Community conditions and rent may establish two stable systems for this population of recruits: good conditions and high rents with household economic insecurities vs. bad conditions and low rents with household economic adequacy and options.

We found a contrast between community and household conditions. The women in the two worst quintiles for community stress reported the least household deprivation. They had economic resources and more human resources, i.e., more people in their households.

One factor that spans the community and household scales is rent. The health area 1990 median rent was strongly negatively associated with the health area chronic community stress index. Rents, thus, largely are determined by community conditions. Rents heavily influence household budgets, especially within a poor, clinic-using population.

The New York City Department of Housing, Preservation, and Development reported that median contract rents increased citywide 5.9% annually between 1993 and 1996 (DHPD, 1999). The median gross rent increased 4.4% annually from 562 to 640 dollars. Residents paid 30% of their incomes for gross rent, as a citywide average. However, Latin Americans paid more than the average: Puerto Ricans, 34.6% and non-Puerto Rican Latinos, 32.1%.

The efficacious community shares and imposes values (Sampson et al., 1997). If the neighborhood community cannot impose its values and bad conditions result, the women can withdraw into their own homes and the homes of their own social network (family and close friends). This also represents a stable, psychosocially acceptable configuration. If community conditions are unstable, not so bad as to drive the decent people out of civic life, but out of control, the people who need a nice neighborhood and to exercise good citizenship experience great distress. The analogy is a cliff. The study participants on top of the cliff are in a stable, safe situation. The study participants at the bottom of the cliff are in a stable, safe situation. But the people in the air between the top and bottom are distressed and frightened.

We mathematically model this latter process in the appendix to this chapter, using a variant of our earlier analyses.

Thus results suggest that the Upper Manhattan study transect is, counterintuitively, for our study participants, a single, integral system which may be inherently unstable, and may fragment under the impetus of sufficient perturbation: Upper Manhattan truly seems to be where the South Bronx was in the early 1970's (Wallace and Wallace, 1998, 2000).

The systematic nature of these relations transcends the different 'ethnic' components of the Upper Manhattan study zone. We conclude that the dynamics of the system are driven by relatively simple indices of deprivation and community stress rather than by inherent qualities of the US social construct of 'race,' although allocation and availability of the resources and services which determine community stress are, under the US system of Apartheid, largely determined at the area level by the perceived 'race' of the inhabitants (e.g., Massey and Denton, 1992).

The Bodymass Index

Besides producing a strong pattern for demoralization scores, the quintile weighted average ICCS produced a strong pattern for average BMI. Epidemiological papers in recent literature on human overweight and stress often report levels of markers of hypothalamus/pituitary/adrenal (HPA)axis activity (example: Fried et al, 2000). Furthermore, experiments on animal models show an unambiguous relationship between adipocyte and HPA functions that 'confirm the inhibitory effect of leptin on the HPA axis response to various stress stimuli' (Giovambatista et al, 2000). Indeed, in sheep, one type of leptin receptor has been localized in the somatostatin-containing neurons of the hypothalamus (Iqbal et al, 2000). Leptin is the so-called 'fat hormone' produced by adipocytes. Thus, both human epidemiological and animal physiological data raise the possibility that the good demoralization scores in the two worst quintiles may be partially paid for with the elevated BMI's. The women in the worst three quintiles have average BMI's above 25, the standard cutoff for overweight.

'Race' and culture may be raised as a possible cause of the differences between the quintiles with respect to the two health outcomes. The lack of significant difference between the demoralization scores and BMI's of Dominican and African-American women and of immigrant and native-born living in the same health areas indicates that 'racial' and cultural differences have at most little bearing.

The women in the three worst quintiles may be forced to spend unusual amounts of time at home or in the homes of friends and relatives because of the neighborhood conditions. The typical activity scenario becomes one of watching television or video movies and eating, either alone or with family and friends, a prescription for overweight/obesity.

Neighborhood conditions such as characterize the three worst quintiles have often been associated with substance abuse. At the health area level, such indicators of substance abuse as cirrhosis and drug death incidences form part of the picture. The women in the study group, however, do not use illegal drugs, nor do they use large quantities of alcohol. Here is where culture may play a role in risk behavior.

The women in the study group may share an important cultural influence on their behavior, although some are Dominican and others African-American: Christian backgrounds. Their upbringing included implicit or explicit tenets with respect to use of drugs and alcohol, especially by women. The approved ways for women to deal with chronic stress in traditional Christian cultures are limited to eating, talking, arts/crafts, and entertainment such as reading and television. These women share a culture that allows comfort foods.

Eating fatty, sugary foods has long been known as a way of coping with stress. In recent years, a large literature has arisen with the discovery of leptin, the so-called fat hormone. Leptin and the adrenal stress hormones relate to each other in complementary ways. For example: leptin and cortisol show complementary circadian cycles with leptin peaking at night and cortisol during the day (Houseknecht et al, 1998). People who do not sleep well because of stress metabolize food differently from those with normal sleep patterns and channel the calories into central abdominal fat deposition (Speigel, Leproult, Van Cauter, 1999). Thus, stress, eating, sleep debt, leptin levels, adrenal hormone levels, weight, and central abdominal fat deposition have been linked.

Our results have implications for the debate on the cause of inequalities in health status among people of different socioeconomic status (SES). On one hand, Wilkinson (1996) and Kawachi et al. (1997) argue that inequality in SES per se fuels this health inequality and that inequality alone imposes stress. On the other hand, Link and Phelan (1999) conclude from their studies that actual deprivations occur due to socioeconomic inequalities. Our results show that the inequalities in the health outcomes (depressive symptoms and overweight/obesity) of the women living in the different quintiles are related to interactions between household economics and neighborhood conditions and are partially based on actual deprivations and insecurities.

11.7 Conclusion

We have taken a preliminary look at the effects of chronic community level social and economic stressors on individual emotional status of a highly functional population of poor mothers as measured by a standard instrument. The graph of quintile averages of the index of chronic community stress plotted against the average demoralization scores of the women in the quintiles of chronic stress yielded an inverted 'U' curve with the maximal average demoralization in the middle quintile.

We offer two possible explanations for this curve, which are not mutually exclusive:

(1) That a particular level of community chronic stress at or about the level of the middle quintile boosts a signal to the individual or her social network to take action which shields her from community stress.

(2) That very good and very bad community conditions lead to two different stable, safe psychosocial configurations with a distressing and dangerous middle area, for this socially-connected cohort of mothers who trust authority enough to enlist in a research study. We speculate that, in either case, this curve depends on the breaking of the weak ties, in the Granovetter (1973) sense, between disjoint social networks and withdrawal from community activity into the homes of members of the social network when community conditions deteriorate. A vicious circle then would ensue, whereby a higher and higher proportion of the civic-minded residents are chased out of the public arena by sheer fatigue, futility, disgust, and fear.

Household economics also influenced demoralization scores. Household economics itself is partly determined by housing costs (rent) which, in turn, are determined by community conditions. Thus, community conditions affected study recruits both directly and indirectly. Women who had experienced a serious deprivation such as recently not being able to pay for food or for housing had much worse demoralization scores than those who did not report such deprivations. Household economics may be one reason for the inverted 'U' shape: in bad neighborhoods, the rents are low and the households have fewer deprivations.

Community conditions determined the pattern of BMI. Where community conditions were bad, the women were decidedly heavier than the women in good neighborhoods. We do not know whether this effect was due to need for 'comfort foods' to cope with the stress or that behavioral pattern of watching television and snacking so common to people staying in the home.

Community conditions and household economics can explain the patterns of health outcomes observed among these women. 'Race', ethnicity, immigrant status, and cultural differences did not explain these patterns.

From the perspectives of the previous chapters, for these study subjects, embedding neighborhood conditions served as general epigenetic catalysts via the more immediate mechanisms of social network configuration, driving the women in the most stressful geographic regions into highly fragmented systems dominated by Granovetter's disjunctive 'strong' social ties. Thus this work suggests the necessity of expanding the epigenetic catalysis concept to a nested hierarchy of structure.

11.8 Chapter Appendix

The essential driving factor in the analysis is the relative strength of 'weak' and 'strong' social ties within the community. Weak ties, in Granovetter's (1973) terminology, are those which do not disjointly partition a community. Meeting and talking with someone regularly in a park might well be a weak tie, while age cohort, ethnicity, and family relationship would be strong ties that indeed disjointly partition a community.

We can envision the culture of a geographically-focused community in terms of a language, in the largest sense, combining behavioral and spoken, and other means of communication between individuals, and between an individual and his or her embedding social structures (R. Wallace and R.G. Wallace, 1998, 1999; Wallace and Fullilove, 1999). Languages are characterized by an information-theoretic 'source uncertainty' which quantifies their efficiency in delivering a message. Languages with high source uncertainty can say much with little, according to theory (e.g., Ash, 1990).

Formally we can express the source uncertainty as follows:

Suppose we can identify a vocabulary of symbols in the language. We ask the number of meaningful statements, $N(n)$ of length n symbols. For regular stationary ergodic information sources, the source uncertainty of the language is given by the usual limiting relation

$$H = \lim_{n \to \infty} \frac{\log N(n)}{n}.$$

(11.2)

Thus statements in the language can be divided into two sets, a relatively small number of high probability which are consonant with the grammar and syntax of the language, and hence meaningful, and a very large number which are not consonant, and are of vanishingly small probability. See Ash (1990), Cover and Thomas (1991), or Khinchin (1957) for details.

As discussed in previous chapters, expressing source uncertainty in this form allows identification of a homology with the free energy density of a physical system that permits importation of techniques from thermodynamics and statistical mechanics into information theory.

The essential point is that we can parametize the source uncertainty characteristic of a geographically focused community in terms of an *inverse* index of its strength of weak ties, that we will call K. If P is, for example, the ensemble average probability of weak ties across the community, then we take $K \equiv 1/P$, and write $H = H(K) = \lim_{n \to \infty} \log[N(n, K)]/n$.

Thus we assume that K is *monotonic in our index of chronic community stress*, that is, K increases uniformly with increase of the index.

Note that, for simplicity, we assume the 'strong' ties to be of a single fixed average probability across the community.

Imposing a thermodynamic formalism, we can write an 'equation of state' for an information system dominated by the relative magnitude of strong and weak social ties as the relation

$$S(K) \equiv H(K) - K dH/dK$$

(11.3)

S is defined as the *disorder* of the system, and the quantity $I = S - H$ is defined as its *instability*.

Wallace and Fullilove (1999) propose that patterns of risk behavior are proportional to the instability I as the average probability of weak ties (or its inverse $K = 1/P$) changes across the community. Here we postulate that, for our study population, which is well embedded in community, not using drugs, not alienated from authority and, in general, strongly connected with family and friends, individual demoralization is proportional to the community instability index I.

Let the function $I(K)$ follow the functional form of the SNR of a stochastic resonance:

$$I(K) = \frac{1}{K^2}[1 + \alpha \exp(1/(2K))]^{-1}.$$

(11.4)

Then, from above, $I(K) = -KdH/dK$ and we can write

$$H(K) = \int -\frac{I(K)}{K} dK =$$

$$\frac{1}{2K^2} - \frac{2}{K}\log[1 + \alpha\exp(1/2K)] - 4\mathtt{Polylog}[2, -\alpha\exp(1/2K)],$$

(11.5)

where we have set the constant of integration to zero. Polylog[x,y] is a standard tabulated function.

An explicit plot shows $H(K)$ is a reverse S-shaped curve, compared to $I(K)$'s inverted 'U'. The richness of trans-community 'language' declines monotonically with increasing K, the inverse of the average probability of weak ties within the community. This is also a way of saying that the capacity of the community as a communication channel declines with increasing K, since elementary information theory arguments show $H(K) \leq C$ where C is the channel capacity.

Note that the decrease of calculated $H(K)$ is most rapid across the 'hump' of the $I(K)$ plot. If the inverse average probability of weak ties across the community, $K = 1/P$, is monotonically determined by the community chronic stress index, this suggests that the functionality of the Upper Manhattan study zone as a social communication channel decreases very sharply indeed with increasing community stress.

We have a paradoxical result applicable to any reverse S-shaped curve, in that instability will be greatest at intermediate points of stress: the 'falling off a cliff' effect.

12

Death at an early age: AIDS and related mortality in New York City

Deborah Wallace, Ph.D.
  Sandro Galea, M.D., Dr.Ph.
  Jennifer Ahern, M.P.H., Ph.D.
  Rodrick Wallace, Ph.D.

12.1 Introduction

Here we present the work of a team associated with the Center for Urban Epidemiologic Studies (CUES) at the New York Academy of Medicine. The principal author was Deborah Wallace, working with Dr. Sandro Galea, at the time, Deputy Director of CUES, and Jennifer Ahern, also then at CUES, who should be regarded as coauthors of the chapter. It focuses on patterns of HIV/AIDS and related deaths as a population/community level disease guild including homicide, drug deaths, and cirrhosis deaths. These data suggest the necessity of expanding the theoretical analysis of the early part of the book from a focus on individual paths of cognitive gene expression driven by epigenetic catalysis to emergent phenomena at the population level, necessarily involving coordinated crosstalk between individuals, subgroups, and perhaps geographic subdivisions.

    We begin by reconsidering the debate on health disparities between populations. These have led to a controversy: are they rooted in actual material deprivations or dependent on socioeconomic inequalities per se? Hypothesizing that the former is the reality, we analyzed mortalities by cause and a variety of socioeconomic, demographic, and environmental data of the 59 community districts of New York City for year 2000. Thus, disparities within a municipality were probed, not between large regions or countries.

    HIV/AIDS death incidence was the most variable of the deaths-by-cause over the community districts. When HIV/AIDS death incidence was regressed against the other health outcomes in stepwise regressions, three other incidences of mortality remained in the regression as statistically significant variables: drug deaths, homicide, and cirrhosis-and-other-liver deaths. We developed a complex index of the four death incidences for the 59 community districts. When it was regressed against a variety of socioeconomic, demographic, and environmental variables in stepwise regressions, the statistically significant variables that remained in the regression were proportion of the households with income below $15,000, unemployment rate, and median income ($R^2 = 0.75$).

The result of the stepwise regression of complex index regressed against socioeconomic, demographic, and environmental factors indicates that economic deprivation is strongly associated with the system of the four highly related health outcomes and accounts for their joint variability over the neighborhoods of New York City. Furthermore, the simple regressions of the complex index and the socioeconomic, demographic, and environmental factors show that low income is much more strongly associated with the index than high income and low educational attainment than high educational attainment. This asymmetry also indicates that actual material deprivation, rather than inequality per se, determines the variability of the mortality index over the community districts of New York City.

Different communities show vastly different patterns of health and of rates of mortality by cause, as has been documented in many reports such as Healthy People 2010 and many published papers and books (Example: Evans, Barer, Marmor, 1994). The underlying cause(s) of the pronounced patterns of poor health and high mortality rates in communities with high prevalence of poverty, low educational attainment, and high proportion of people of color is a matter of scientific controversy. Wilkinson has focused on income inequalities per se (1996), whereas Link and Phelan (1999) have documented actual deprivations of potentially adverse impacts on health and mortality rates such as access to Pap tests.

Wilkinson put it thus (p. 3, 1996): 'Indeed, the evidence suggests that what matters within societies is not so much the direct health effects of absolute material living standards so much as the effect of social relativities.' This quote clearly delineates the debate.

Hillemeier et al. (2003) explored whether relative or absolute standards for child poverty were associated with infant and child mortality. Poverty defined by national standards was strongly related to mortality among infants and children, but the state's relative position in the economic hierarchy was not. The authors concluded that the major economic influence was the 'families' capacity for meeting basic needs.' Ross et al. (2000) compared and contrasted the effects of income inequality in Canada and in the USA. Although the combined analysis of populations of metropolitan areas showed that income inequality was a statistically significant explanation for mortality in all age groups except the elderly, when only Canadian data were analyzed, no significant association was found between income inequality and mortality. Thus, income inequality per se may not automatically determine mortality rates. The controversy may be framed as relative income vs absolute, or as psychosocial vs material determinants.

Although Wilkinson posits relative socioeconomic status or position as the basic influence, his comparisons of the UK and Sweden with respect to infant mortality and male life expectancy show steep gradients in the UK and little or no pattern in Sweden where basic necessities are provided by social spending (pp. 87-88, 1996).

Our hypothesis is that material deprivation and absolute income levels strongly influence the geographic prevalence and incidence patterns of many health outcomes. If our hypothesis is correct, then the prevalences and incidences of those health outcomes with the greatest inter-neighborhood variabilities will show strong statistical associations with measures of neighborhood economic deprivation such as proportion of low income households and percent of adults who are unemployed.

Furthermore, measures of neighborhood economic wellbeing such as percent of high income households will have weaker statistical associations with incidence of morbidity and mortality than do the measures of economic deprivation because such factors as percent of households with high income and percent of adults with high educational attainment do not impose a direct depriving stress on a neighborhood, nor does their presence lift deprivation burdens directly from a largely poor neighborhood. Only politically important neighborhood concentrations of high-income households and adults with high educational attainment lift some burdens of material deprivation by ensuring favorable public policies and private practices. Thus, we expect a threshold effect of concentrations of high- income households and high educational attainment adults on incidence and prevalence of morbidity and mortality, a non-linear negative relationship. Conversely, we expect a positive linear relationship between morbidity/mortality and concentration of low-income households and low educational attainment adults.

Here we examine patterns of HIV/AIDS death rates of the 59 community districts (CDs) of New York City and of other associated health outcomes. HIV/AIDS death rate was chosen as the focal health outcome because it is the most variable among the major causes of death in New York City. We shall find the health outcome incidences that form a single system of disease/death with HIV/AIDS death. We then shall determine the socioeconomic, environmental, and demographic factors associated with this system at the CD level. We shall develop a socioeconomic, environmental, and demographic model that most parsimoniously explains the joint pattern of HIV/AIDS death rates and the rates of death closely associated with HIV/AIDS death rates. Such a model and the statistical associations found in the course of its development would provide evidence in the controversy over whether actual material deprivation or relative economic status determines the well-documented municipal patterns of health disparity.

12.2 Data and analysis

The data for the 59 CDs consists of a wide variety of demographic, socioeconomic, environmental, and health variables. These variables are listed in the appendix. Data on population, ethnicity, age, educational attainment, employment/unemployment, poverty, and household income were taken from the 2000 Census, as reported by the New York City Department of Planning in its online '2000 Census Community District Demographic Tables' and '2000 Census Community District Socioeconomic Tables'.

The environmental variables such as structural fires and percent clean streets were taken from the online database Mayors Office of Operations Neighborhood Statistics. Number of housing violations per housing unit and number of housing units were acquired from the New York City Housing and Vacancy Survey for 1991 and 1999.

From the New York City Department of Health's online table, Summary of Vital Statistics 2000: the City of New York, came the health outcome data: total deaths per 1000 and deaths by cause per 100,000 population, and live births and birth outcome characteristics. See the Chapter Appendix for a list.

All statistical analyses were performed with Statgraphics Plus, version 7. Standard statistical methods were used for data analysis: statistical descriptions (mean, standard deviation, median, maximum/minimum), simple regressions, t-test, and backwards stepwise regressions.

This study featured two sections: the first to find the most geographically variable health outcome and to find any other health outcomes that form a single system with it and the second to find the socioeconomic, demographic, and environmental factors most closely associated with these highly variable health outcomes. The most variable outcome turned out to be HIV/AIDS death incidence.

Simple regressions were performed between HIV/AIDS death incidence and those of the other health outcomes. For those health outcomes that reached an R^2 of at least 0.4 in these simple regressions, stepwise regression was used to find the health outcomes that together formed statistically significant variables when regressed against HIV/AIDS death rate. The health variables were regressed five at a time against HIV/AIDS death rate, until the most parsimonious model was found. Only five at a time were used to assure stability in the stepwise regression for 59 CDs.

A complex health outcome index was developed from HIV/AIDS death incidence and the health variables associated with it in the stepwise regression. The CD incidence of each component was divided by the relevant median of all 59 CDs. The median-standardized incidences were then summed to form the complex health outcome index. Socioeconomic, demographic, and environmental variables were regressed against this complex index. The form and strength of the associations were noted because these attributes can provide evidence as to the determinant of the observed variability over the CDs (inequality vs absolute deprivation). Those contextual variables with R^2 above

0.4 were regressed five at a time against the complex index in backwards stepwise regression until the factors forming the most parsimonious model were found.

T-tests probed whether division of the CDs into those with high and those with low complex health outcome index produced significantly different means of socioeconomic, demographic, and environmental factors. By 'high', we mean an index above the sum that would be obtained if each component of the complex index were unity (all incidences of the CD were the medians); by 'low', we mean an index below that sum.

One ecosystem measure was used, the Shannon-Weaver biodiversity index, as a metric of CD diversity:

$$SW = -\sum_i P_i \log[P_i]$$

where P_i=the proportion of the community comprised of a species (here we looked at diversity of 'species' of ethnicity and income, separately). Ethnicities were the standard Census classifications, and income classifications are displayed in appendix A. The Shannon-Weaver diversity index has three properties:

1) For a given number of species, SW takes its maximum value when all species are present in equal proportions (i.e., are completely even).

2) Given two completely even communities, the one with the larger number of species has the greater value of SW.

3) SW can be split into components that are additive (Pielou, 1977).

Thus, when we calculate Shannon-Weaver diversity, we are measuring how well a community is mixed. We measure ethnic and economic integration.

12.3 Results

Basic descriptive statistics for incidences per 100,000 of selected top causes of death in New York City show diverse patterns (Table 1). The most common, cardiovascular and other heart disease (CVDH) and cancer, show the least variability. HIV/AIDS death rates show the greatest variability over the CDs, whether measured by the standard deviation compared with the mean or by the ratio between the maximum and minimum. Although CVDH death rates and cancer death rates were much greater than those of HIV/AIDS, the maximal HIV/AIDS death rate was much greater than the maximal rates of other causes with similar means such as flu or stroke. The minimal HIV/AIDS death rate was, on the other hand, much below those of other causes with similar means. Thus, this particular cause of death exhibits a remarkable variability in incidence.

Table 2 displays the health outcomes that are associated with HIV/AIDS death rate in simple regressions with an R^2 above 0.2. The first three outcomes, cirrhosis and other liver deaths, drug death, and homicide incidence,

Table 1. Summary Statistics of Selected Death Rates by Cause: 59 Community Districts rates per 100,000 in 2000

Cause	mean	median	SD	Minimum	Maximum	Max/Min
HIV/AIDS	26.4	17.2	23.2	0.9	99.2	110.2
flu	27.3	25.4	12.5	7	56.5	8.1
diabetes	23.4	21.2	11.8	2.9	63.4	21.9
stroke	23.3	21.7	6.7	12.8	42	3.1
heart	290.9	262.9	121.3	113.3	700.2	6.18
cancer	157.3	150	38.2	87.4	259.1	3
homicide	9.1	8.1	6.5	0.8	24.8	31
drugs	11.1	9.6	7.3	2.1	35	16.7
liver	7.1	6.4	4.3	1.7	23.3	13.7
chronic respiratory	19.3	17.8	7.4	5.8	43.3	7.5
accident	11.5	11.2	3.4	4.5	18.7	4.2
suicide	5.1	5.1	2.5	1.2	12.6	10.5
AIDS incidence in 1999 (not death rate)	70.2	72	45	9	193	21.4

are markers of risk behaviors of violence and substance abuse, whereas births to teenagers and births to single mothers indicate early sexual activity and sexual activity outside of marriage. Diabetes is a marker of a different type of risk behavior: obesity-causing eating and inactivity behaviors.

Table 2. Other Health Outcomes Associated with HIV/AIDS Death Incidence
Results of Simple Regressions of HIV/AIDS Death Incidence Against Health Outcomes
Only results with R-sq over 0.2 are displayed.

Health Outcome	R-sq	F-ratio
liver death incidence	0.47	49.44
drug death incidence	0.58	80.04
homicide incidence	0.57	72.48
diabetes death incidence	0.46	48.56
low-weight birth rate	0.37	33.37
percent births to teens	0.53	63.21
percent births to single mothers	0.45	46.71

All associations are positive and linear. All P's are less than 0.00001.

When these independent variables are regressed against HIV/AIDS death rate in stepwise regression, the most parsimonious model which results includes cirrhosis and other liver deaths rate, drug death rate, and homicide death rate (Table 3). These three associated independent variables are grouped with HIV/AIDS death rate in a complex index.

We shall call the complex variable HHLD: HIV/AIDS, Homicide, Liver, and Drugs. HHLD for each CD is the sum of the CD's four death rates divided

Table 3. Results of Stepwise Regression of HIVAIDS Death Incidence vs Associated Health Outcomes

Most parsimonious models

independent variable	coefficient	F-remove	total F	R-sq
liver death incidence	1.28	4.46		
drug death incidence	0.88	5.31		
homicide incidence	1.8	31.61		
constant	-9.33		53.27	0.74

Heart death incidence was significantly associated negatively with HIVAIDS death incidence, but the R-sq was less than 0.2. The following model shows an interesting relationship in the stepwise regression, however. The other variables did not "wash out" heart from the regression.

independent variable	coefficient	F-remove	total F	R-sq
liver death incidence	1.2	4.46		
drug death incidence	0.88	5.8		
homicide incidence	1.59	25.6		
heart death incidence	-0.04	6.65		
constant	3.74		45.95	0.77

by their respective medians over the 59 CDS. Because of skewing in some variables, the median, rather than the mean, was selected for standardizing. Several socioeconomic variables are strongly associated with HHLD: median income, proportion of households with low income, unemployment rate, and income diversity as measured by the Shannon-Weaver diversity index (Table 4).

Table 4. Simple Regression Results of HHLD vs SES Variable

HHLD=(HIVAIDS/median)+(Homicide/median)+(Liver/median)+(Drugs/median)
median=each variable's median over the 59 CDs

Results of simple regressions with R-sq over 0.25

independent variable	form of assn	R-sq	P
% adults without high school diploma	linear positive	0.41	<0.00001
% adults with college degree or more	non-lin. neg.	0.31	0
% of population over 65 years of age	non-lin. neg.	0.34	<0.00001
median income	non-lin. neg.	0.42	<0.00001
% households with less than $15K income	linear positive	0.65	<0.00001
% households with more than $150K income	non-lin neg	0.25	0
black/totalpop	non-lin pos	0.44	<0.00001
white/totalpop	non-lin neg	0.48	<0.00001
(black+hispanic)/totalpop	non-lin pos	0.56	<0.00001
% births to teenagers	linear positive	0.64	<0.00001
% births to single mothers	non-lin pos	0.62	<0.00001
% adults unemployed	linear positive	0.68	<0.00001
income diversity (Shannon-Weaver)	linear negative	0.50	<0.00001

Percent of the adult population without a high school degree is positively linearly associated with HHLD ($R^2 = 0.42$), whereas percent of the population with more than a college education is negatively non-linearly associated ($R^2 = 0.30$). Percent of households with low income is positively linearly associated with HHLD ($R^2=0.66$), whereas percent of households with high income is negatively non-linearly associated ($R^2 = 0.24$). Ethnic diversity was significantly associated with HHLD but with a low r-sq. Unemployment rate showed the highest R^2 (0.68).

When backwards stepwise regression (Table 5) was used to develop the parsimonious model of potentially explanatory variables for the pattern of HHLD over the 59 CDs, three variables were selected by the regression: unemployment rate, median income, and proportion of households with less than $15,000 income ('low income' in this paper). All ethnic variables dropped out of the model, as did proportion of births to teenage mothers, proportion of births to single mothers, and income diversity. Median income in this model is associated positively with HHLD, although in the simple regression of median income as the sole independent variable, the association was negative.

Table 5. Stepwise Regression Results: HHLD vs Variables with R-sq Over 0.4

independent variables	coefficient	R-sq	F	P
unemp/(totalpop-under18)	55.19			
median income	0			
proportion low income households	20.8			
constant	-6.18			
		0.73	52.42	<0.00001

HHLD of a CD would sum to 4 if all four components of that CD were at the median incidence of all 59 CDs. When the CDs were divided into those with HHLD above 4 and those with HHLD below 4, 29 were in the first division (high HHLD) and 30 in the second (low HHLD).

Table 6 compares selected socioeconomic and environmental variables in the two groups of CDs. Ethnic diversity showed only a trend to difference between the two groups, and population density no significant difference. All other variables showed statistically significant differences between the means of the two groups of CDs: percent black, percent Hispanic, educational attainment, percent of households with low income, percent of households with high income, median income, unemployment rate, income diversity, structural fires per population unit, percent of clean streets, and percent of housing units with three or more serious violations. Particularly large differences in means appeared in the following variables: percent black (38.95% vs 11.20%), percent of households with low income (32.9% vs 17.5%), percent of households with high income (3.4% vs 6.9%), fires per 1000 people (5.3 vs 2.6), and unemployment rate (8.08% vs 4.33%).

The map of Figure 12.1 displays the CDs at the extreme high end of the HHLD range (6 or more) and at the extreme low end (2 or less).

Table 6. High vs Low HHLD CD's: Socioeconomic and Service Indicators

	mean of high	mean of low	t	P
% black	38.95	11.2	5.01	0
% hispanic	35.3	21	2.78	0.01
(educ1+educ2)/pop25	0.37	0.23	4.81	0
(educ6+educ7)/pop25	0.2	0.32	-2.67	0.01
(inc1+inc2)/hholds	0.33	0.18	7.15	<0.000001
(inc9+inc10)/hholds	0.03	0.07	-2.21	0.03
medinc	29481.6	47638.3	-5.4	0
incdiverse	2.03	2.14	-4.55	0
ethdiverse	0.4	0.45	-1.82	0.08
fires/1000 people	5.3	2.6	3.69	0
cleanst	75.2	84.6	-5.12	0
popdensity	46448.6	38898.8	-1.17	0.25
mnydef99/ohunit99	0.39	0.3	3.87	0
unemp/(totalpop-under18)	8.08%	4.33%	8.1	<0.0000001

ethdiverse=ethnic diversity measured by Shannon-Weaver diversity index
incdiverse=income diversity measured by Shannon-Weaver diversity index
mnydef99/ohunit99=proportion of housing units with 3 or more serious violations in 1999

Table 7 displays the median incomes and percent low income households of CDs with very high HHLD (over 6) and very low HHLD (less than 2). Two CDs in the former group with median incomes over $40,000 are gentrifying neighborhoods that had been part of the Brooklyn poverty belt twenty years ago: Park Slope and Williamsburgh. One CD in Manhattan (Chelsea/Clinton/Hells Kitchen) also falls into this pattern in having a median income above $40,000 and having been part of a poverty zone in the past. On the other hand, the two CDs in the latter group with percent low income households over twenty (Brooklyn CDs 12 and 15) are areas into which low income households had fled relatively recently. The median incomes of these two CDs fall well below the average of the very low HHLD group of ten CDs.

The deaths from HIV/AIDS, liver disease, drug use, and homicide impose greatly different intensities of public health problem to these two sets of communities. In the high HHLD CDs, HIV/AIDS deaths comprised as high as 14% of the total deaths in 2000 (Table 8). The highest percent of deaths from HIV/AIDS in the low HHLD group was 1.5%, only a bit more than one-tenth that in its high HHLD analog. Deaths from the four related causes ranged from 8 to 24% among the high HHLD CDs and from 1 to 3.5% in the low group. Among the high HHLD CDs, deaths from the four related causes are a significant public health problem.

Diabetes death incidences were also highly associated with HIV/AIDS deaths in the simple regressions, but were so collinear with the other three related deaths that the diabetes deaths were omitted from the parsimonious

12 Death at an early age: AIDS and related mortality in New York City

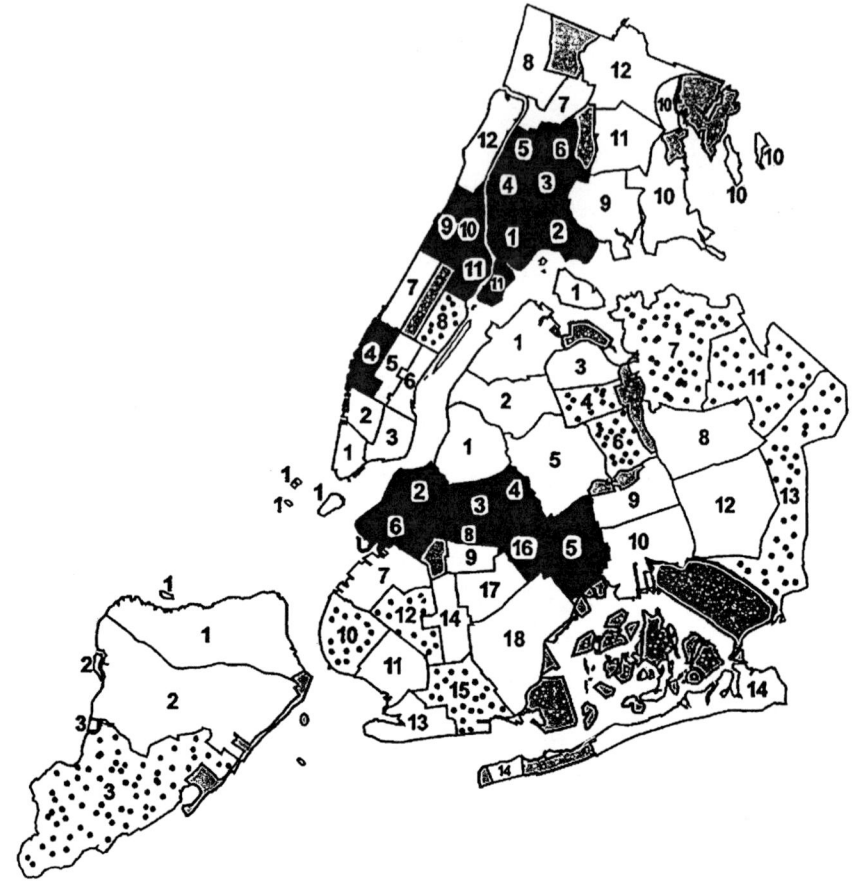

Fig. 12.1. Map of Community Districts with Very High and Very Low HHLD. Very High = 6 or more (black filled). Very Low = 2 or less (stippled).

model in the stepwise regression. Among the high HHLD CDs, the contribution made to total deaths by diabetes ranged from 2.6 to 9.8%; among the low HHLD CDs, it was 1.3 to 3.0% (Table 8). In contradiction to the expectation that diabetes deaths are an index of the population of the elderly, they are inversely associated with the percent of the population that is over 65 years of age. Among the high HHLD CD's, those over 65 years old contribute 5.1 to 11.6% of the population (average 8.5%); among the low HHLD CDs, they are 8.3-18.8% of the population (average 14.4%) (Table 8).

Table 7. Economic Characteristics of CDs with very High and very Low HHLD

	High HHLD Group				Low HHLD Group		
CD	%below $15K	%above $75K	medinc	CD	%below $15K	%above $75K	medinc
Bx 1	48.21	5.21	16,000	Bk 10	18.46	24.54	43090
Bx 2	43.51	5.22	17,130	BK 12	29.39	15.89	29780
Bx 3	47.29	5.69	16600	Bk 15	24.18	21.69	37450
Bx 4	40.42	5.76	21275	Mn 8	9.54	49.42	74130
Bx 5	41.32	6.60	20620	Q 4	19.93	16.73	36470
Bx 6	47.40	5.05	16530	Q 6	16.66	27.60	47520
Bk 2	20.98	28.15	44180	Q 7	16.29	24.58	43480
Bk 3	38.61	10.52	23495	Q 11	9.90	35.54	57960
Bk 4	38.25	7.13	22100	Q13	9.17	33.71	57080
Bk 5	35.08	10.25	25505	SI 3	9.07	40.55	64260
Bk 6	16.37	33.53	53090				
Bk 8	30.16	13.63	28780				
Bk 16	44.79	7.91	18750				
Mn 4	17.31	33.40	50580				
Mn 9	32.59	15.40	27365				
Mn 10	41.72	8.33	19920				
Mn 11	40.65	9.75	21295				

12.4 Discussion

Our results support the hypothesis that absolute income and material deprivation strongly influence the New York City municipal pattern of unequal incidences of HIV/AIDS deaths, drug deaths, liver deaths, and homicide. Let us briefly review the results. HIV/AIDS death incidence is the most variable of the major death incidences by cause, the highest being 110 times that of the lowest. Many other health outcomes were statistically associated with HIV/AIDS death rate in simple regressions but only three proved significantly positively associated with HIV/AIDS death rate in stepwise regression: homicide rate and incidences of liver death and drug death. A complex index of this system of four deaths-by-cause developed by summing each CD's rates standardized by the respective medians became the dependent variable in simple regressions with socioeconomic, demographic, and environmental factors as the independent variables.

Although both high and low SES factors were statistically significantly associated with HHLD, the low SES factors were usually associated linearly and more strongly than the high SES factors which were non-linearly associated with HHLD. In stepwise regression, three factors explained three-quarters of the pattern of HHLD over the CDs: the percent of households in the two lowest income classes, median income, and unemployment rate. Division of the CDs into those with HHLD above and below 4 showed significant differences between the two groups with respect to socioeconomic, demographic, and environmental factors. Even the number of structural fires per unit population and the proportion of clean streets differed significantly.

What focused our attention on HIV/AIDS death incidence is that it is the most variable of the major death incidences by cause. The possible reasons for the great disparities between neighborhoods are several. HIV/AIDS death incidence marks the endpoint of many processes from the incidence of HIV

12 Death at an early age: AIDS and related mortality in New York City

Table 8. Percent of Total Deaths from Four Related Causes
Very High HHLD CDs vs Very Low HHLD CDs

High

CD	%HIV/AIDS	%Liver	%Drugs	%Homicide	Total	%diabetes	%>65 yrs*
Bx 1	9.74	2.62	2.43	1.69	16.48	5.62	7.7
Bx 2	6.64	2.8	5.59	2.8	17.83	5.94	6.9
Bx 3	13.23	3.11	4.67	3.31	24.32	5.26	6.7
Bx 4	8.67	1.62	3.58	2.43	16.3	5.66	6.9
Bx 5	14.12	1.62	1.79	3.73	21.26	4.87	5.1
Bx 6	11.38	0.81	3.05	3.05	18.29	9.76	7.7
Bk 2	5.84	1.23	1.78	1.4	10.25	5.08	10.1
Bk 3	8.21	1.59	2.26	2.18	14.24	5.11	9.1
Bk 4	8.07	1.68	3.03	2.86	15.64	3.87	6.4
Bk 5	5.96	1.12	2.05	2.61	11.74	4.19	7.9
Bk 6	5	1.92	1.79	0.64	9.35	5.26	8.9
Bk 8	6.56	1.14	1.85	2.43	11.98	5.15	9.4
Bk 16	9.9	0.99	3.3	1.67	15.86	4.62	7.6
Man 4	7.51	1.88	2.35	0.63	12.37	2.66	11.6
Man 9	5.87	1.2	2.67	2.8	12.54	3.87	9.9
Man 10	6.49	1.16	2.4	1.87	11.92	4	11.5
Man 11	3.79	0.9	2.17	1.17	8.03	4.61	11.5
Mean of total					14.61	5.03	8.52

Low

CD	%HIV/AIDS	%Liver	%Drugs	%Homicide	Total	%diabetes	%>65 yrs*
Bk 10	0.53	0.53	0.89	0.09	2.04	1.33	16.2
Bk 12	0.57	0.33	0.41	0.41	1.72	1.8	12.6
Bk 15	0.27	0.4	0.8	0.34	1.81	1.94	18.1
Mn 8	1.08	0.41	0.34	0.27	2.1	1.69	14.2
Q 4	1.54	0.61	0.61	0.31	3.07	1.69	8.3
Q 6	0.2	0.2	0.39	0.2	0.99	2.94	18.8
Q 7	0.65	0.6	0.65	0.49	2.39	1.68	15.8
Q 11	0.14	0.41	0.55	0.14	1.24	2.05	17.1
Q 13	1.2	0.47	0.47	1.42	3.56	3.02	12.1
SI 3	0.56	0.68	0.68	0.23	2.15	1.91	10.4
Mean of total					2.11	2.01	14.36

*%>65 yrs is the percent of the CD population over 65 years, not the % of deaths contributed by those over 65 years.

infection, through access to and use of HIV testing, access to and use of anti-viral therapy, the underlying health of the CD population, adherence to the difficult treatment regime, and sociogeographically influenced factors that lead to drug resistance of the virus. HIV/AIDS death incidence is, perhaps, the most sensitive indicator of health inequality.

McFarland et al. (2003) noted that even after HAART was introduced, neighborhood income predicted survival of HIV infected people. Indeed, only after introduction of HAART did significant variability in survival appear between neighborhoods of different income levels. Also, del Rio and Barragen (2004) reported that introduction of HAART did not affect incidence of or mortality from AIDS in inner-city populations of Atlanta. R.G. Wallace (2002) documented how slowly AIDS incidence declined in poor ZIP codes of New York City after introduction of HAART and how some populations still showed no change in incidence years after incidences in wealthy areas began steep decline.

Yet, HIV/AIDS death incidence is not a lone, isolated example of large inequalities over the CDs of New York City. Many other health outcomes showed strong associations with HIV/AIDS death incidence. Although these other health outcomes were not as variable as HIV/AIDS death incidence over the 59 CDs, they showed distinct unevenness (Table 1). These outcomes tended to be those already noted in the public health literature as occurring most densely in poor urban neighborhoods: low-weight births, diabetes deaths, liver deaths, drug deaths, homicide, infant mortality, births to single mothers, and births to teenagers. What is new in this study is the discovery of a system of three outcomes, that are clearly strongly linked ($R^2 = 0.75$ in stepwise regression). These four causes of death contributed an average of almost one-sixth of the total deaths over all the very high HHLD CDs and over one-fifth of the total deaths in two of the very high HHLD CDs. They are, thus, a major public health problem in these neighborhoods and take away a disproportionate total of years of life lost below age 65 because they are concentrated among the young and middle-aged adults, rather than the elderly. In these neighborhoods, the proportion of the population over age 65 averages only 8.5% and gets as low as 5.1%. These are the neighborhoods of death at an early age.

Ecologists would call the four health outcomes in the single system a 'species guild', entities that regularly appear together. The significance of species guilds is that the components thrive in the same environment. Clearly, species of animals that appear together consistently fit into the same general environment. The four members of the health outcomes 'guild', likewise, share a particular neighborhood environment which is defined by the results of the stepwise regression: high proportion of low income households, high unemployment rate, and (counterintuitively) high median income in the context of widespread low income and high unemployment. The shared environment of high HHLD is economic deprivation and lack of economic opportunity.

As Table 7 shows, some CDs with very high HHLD had over 15% of the households with income below $15,000, over a quarter of the households with incomes over $75,000, and median incomes well above $40,000. High median income in the context of large pockets of low income and high unemployment rate potentially has two impacts: increased strain on low income households through increased costs of living and destabilizing of the neighborhood social structure as poor families move out and richer households move in. This process is gentrification. These impacts may interact synergistically so that social capital declines as the adequacy of household income declines. Gentrification is one form of residential destabilization. Homicide (and other violent crime) is a marker of low neighborhood efficacy, in the sense of Sampson et al. (1997), as is substance abuse (Wallace, 1990). These studies noted that neighborhood efficacy depends on residential stability. The three health outcomes that form the 'guild' with HIV/AIDS death are markers of community disruption and social network fragmentation, as well as economic deprivation. Thus, gentrification provides an important example of how large health disparities are generated at the municipal level.

In the simple regressions, low household income and educational attainment are linearly positively associated with HHLD (have no threshold in their effect on HHLD), but high household income and educational attainment are non-linear, i.e., have a threshold, and show much weaker effects, as shown by the lower R^2. This means that high income and educational attainment must attain a certain concentration in the population before they affect HHLD, whereas low income and educational attainment exert an influence over HHLD at all concentrations in the population. This result is consonant with our hypothesis and its corollaries, namely that high income households and high educational attainment adults do not impose a direct depriving stress and must attain a threshold concentration in a neighborhood before they can exert enough political and economic influence to lift burdens of stressful public policies and private practices.

If the health inequality over the CDs is rooted in economic or educational inequality per se, median income should be negatively linearly associated with HHLD, as Wilkinson (1996) showed the linear effect of national median income on overall mortality rate. Differences at the high end should be equally effective as differences at the low end, if inequality per se is the driving force in health disparity. Likewise and for the same reason, proportion of high income households and percent of high educational attainment adults should have an equal and symmetric effect on HHLD as proportion of low income households and percent of low educational attainment adults. The effects of proportion of very high income households may arguably be less than those of proportion of very low income households because the former is much lower than the latter.

However, the percent of adults with high educational attainment is not very different from the percent with low educational attainment over the whole city: in high HHLD CDs, 20% of the adults are classed in the top two educational attainment groups; in the low, 32%. In the high HHLD CDs, 36.5%

of adults are classed in the two lowest groups; in the low CDs, 22.5%. The difference in the shape of the relationship between HHLD and the proportions of adults with differing educational attainment indicates a difference in the effects of high and low educational attainment in the CD populations, rather than a difference in order of magnitude of presence of the two educational levels. Daly et al. (1998) found similarly that 'inequality measures reflecting depth of poverty show stronger correlations with mortality than do inequality measures reflecting heights of affluence.'

Median income exhibits a threshold above which effects on HHLD are muted, whereas unemployment rate has no such threshold. The influences of these two variables on HHLD further indicate that inequality per se cannot cause the pattern of HHLD among the 59 CDs of New York City; rather, the disease guild feeding into HHLD arises out of true deprivations that, in turn, arise from income below the threshold and from unemployment.

The stepwise regression of HHLD as the dependent variable and socioeconomic and environmental factors as independent variables yielded a result that confirms the conclusion of deprivation causality. The two most powerful explanatory variables in this regression, unemployment rate and proportion of households that have low incomes, are indices of economic deprivation. Median income also enters this regression with the implication that CDs with high median income and high unemployment rate and/or high proportion of households with low income show high HHLD. This is precisely the paradoxical economic setting of gentrifying neighborhoods.

The high HHLD of the three gentrifying CDs cannot be attributed to the presence of a gay community in each because other CDs with large gay communities (Manhattan CDs 2 and 7) are not part of the very high HHLD group. The influential factors in the very high and very low HHDL groups appear to be both economic and social. Historic socioeconomic processes that render a community either vulnerable to high HHLD or relatively immune to HHLD appear to function for long periods of time even after major changes; indeed, they may be temporarily reinforced by the deprivation stresses on poor households from gentrifying changes in the high HHLD CDs. The high HHLD CDs do not exist as isolated neighborhoods, but as distinct large blocks of neighborhoods (figure 12.1). The two gentrifying CDs with high HHLD are obviously part of the same band of neighborhoods as the poor CDs with high HHLD, the old poverty belt of Brooklyn. Harlem (Manhattan CDs 9,10,11) and the South Bronx (Bronx CDs 1-6) form the other blocks of high HHLD CDs.

Galea et al. (2003) showed that people in poor neighborhoods had a higher risk of becoming drug users than those in other neighborhoods and that drug users in poor neighborhoods had a higher risk of dying of overdose. R.G. Wallace (2003) showed that use of anti-viral therapy was much slower to penetrate poor neighborhoods than others and that poor neighborhoods continued to show high incidence of AIDS long after rates in other neighborhoods declined; R Wallace (1990)had already shown that AIDS incidence (during the time when it could serve as an indicator of HIV infection incidence) was much higher in poor destabilized neighborhoods than others. D. and R. Wallace (1998) had described the Bronx homicide patterns of 1970, 1980, and 1990 and shown that violent death incidence was highly associated with such factors as unemployment rate. Liver cirrhosis death incidence was also proven to be highly concentrated in poor neighborhoods and associated with neighborhood conditions (Wallace and Wallace, 1990).

Thus, historically, the four variables that comprise HHLD have been concentrated in poor neighborhoods for twenty years. Although the aggressive anti-viral therapy has drastically decreased AIDS incidence and death rate in all neighborhoods, it has not eliminated the disparities. Indeed, the variability among the CDs for HIV/AIDS death rate is greater than that of the other major causes of death and greater than the variability of AIDS incidence. Our analysis indicates that patterns of death from these four related causes have roots in economic and other deprivations.

The breadth and intensity of deprivation and insecurity of working class and poor households in New York requires a couple of examples to inform readers who may live in other societies or even other more progressive US cities. In 2003, one in five New York City residents and one in four children used soup kitchens and/or food 'pantries' (charitable sources of food to be prepared and eaten at home) (Office of the Public Advocate, 2003). More than 75% of metropolitan regional households with income below $20,000 pay more than 30% of that income on housing (Perrotta et al., 2004).

Other authors over the past thirty years have documented similar patterns for drug deaths, alcohol/liver deaths, and homicide and linked these deaths to economic differences between neighborhoods and to power relationships that hinge on the economic differences. Eyer (1977) found that cycles of unemployment determined death rates and that variations in drug and alcohol deaths together explained 27% of the variation in death rates with changes in the economic cycle; but the greatest part of the mortality variation (72%) was explained by changes in social relationships.

Rodriguez-Sanz et al. (2003) examined premature mortality across 17 autonomous communities of Spain with respect to power relations, labor market, income inequality, absolute income, poverty, civil associations, and welfare state variables. Premature mortality decreased somewhat more in socialist than in conservative communities for total mortality as well as specific causes such as cirrhosis. Several of the other variables such as labor market, welfare state, and income also were significantly associated with the health indicators.

Our data (Table 8) show that deaths from the four related causes contribute significantly to the total number of deaths in the high HHLD community districts. This is not just a matter of high variability but low numbers. Diabetes deaths are also major contributors in these CDs to the total number of deaths. These are the communities in which a low percent of the population achieves age 65 or more (average 8.5%). The famous paper by McCord and Freeman which showed that life expectancy in Central Harlem (Manhattan

CD 10) was lower than that of Bengladesh enumerated the conditions that harvested the young and were important sources of the startlingly diminished life expectancy: AIDS, homicide, drugs, and cirrhosis along with stroke (McCord and Freeman, 1990).

Our data are cross-sectional and cannot reflect the historic processes that led to the revealed patterns. Thus, we can conclude that patterns of mortality from the four 'guild' members appear to be determined by absolute economic deprivation, but we cannot ascribe that deprivation solely to the economic factors that enter the stepwise regression because we do not know to what extent this deprivation afflicts the communities because of decreased social capital due to community destabilization. Presence of rapidly gentrifying CDs among those with very high HHLD hints that destabilization blunted community efficacy, as Sampson et al. (1997) described.

Thus, the impact of concentrated low income households and high unemployment rates on these mortality rates may be magnified by the loss of social structure in destabilized CDs. Poverty that is not buffered by help from old friends and neighbors may impose a more serious absolute deprivation than the same poverty in the context of old, stable neighborhoods with dense layers of social networks. Answering this question would require a combination of longitudinal and cross-sectional studies, i.e., the ecologist's classical 'spatiotemporal' analysis.

An objection may be made that we have structured a false dichotomy between deprivation/poverty and relative socioeconomic status or position. This objection ignores a debate that has raged in the scientific discourse on socioeconomic determinants of health and health inequality. The debates involves precisely this dichotomy. For example, Wilkinson framed it thus: 'The importance of income distribution implies that we must explain the effect of low income on health through its social meanings and implications for social position rather than through the direct physical effects which material circumstances might have independently of their social connotations in any particular society' (p. 176, Wilkinson, 1996). Whether actual deprivations and necessity of addressing realistically perceived threats to procuring basics are the major determinants of health patterns or whether ranking in SES is has profound policy implications for addressing the problem. Even Swedish sociomedical scientists have opted for policy that includes reduction of poverty rates, better integration of the labor market, and reductions in occupational exposures (Ostlin, Diderichsen, 2001), an indication that even in Sweden where deprivation is least prevalent of all the well-studied societies, differences in basic material resources, in the effort to obtain basic necessities, and in chemical exposures, may be thought to contribute to the detected health inequalities.

Outside of Sweden, in such societies as the UK and the USA, actual deprivation has been widely acknowledged as a major influence on health inequalities for a long time. For example: 'Overall, in Britain around 50% of the health disadvantage of lone mothers is accounted for by the mediating factors of poverty and joblessness... .' (Whitehead et al., 1997-1998).

This chapter has analyzed geographic distributions of health outcomes and their potential determinants among the neighborhoods of a large American city. In a sense, it updates and reconfirms the classic study of Central Harlem life expectancy and causes of death by McCord and Freeman (1990). However, it avoids the error of scale of that study: neighborhoods are compared only with neighborhoods and not with an entire country such as Bangladesh.

At this scale of neighborhood and groups of neighborhoods, the processes determining differences in health between these entities may differ from those at larger scales such as regions and nations. Indeed, poverty rate was a determining factor in county patterns of AIDS incidence within metropolitan regions but loss of manufacturing jobs was the determinant in patterns of AIDS incidence within the network of metropolitan regions (Wallace et al., 1999). So caution must be used not to apply these results to a different scale or level of organization. Comparison with London, Stockholm, or any other major city is appropriate. Comparison with whole nations as is common in international health research is not appropriate. Indeed, one of the grave errors in much discourse on health inequalities may be that of scale and of organizational level.

A particular problem of scale, from the perspectives of this book, is to enlarge a formal analysis that has focused almost entirely on individual development and its pathologies to include the emergent properties inherent to population and community patterns of chronic disease and developmental disorder in which individuals and subgroups may reinforce behaviors or become enmeshed in more subtle, possibly highly punctuated, crosslinkages akin to phase transitions. This will require a nontrivial extension of theory. The observation that infectious disease, behavioral pathology, and addictions can become a disease guild driven by population-level stressors implies that individual level epigenetic catalysis is greatly complicated by cross-scale interactions. These may include collective, rather than individual, phenomena of gene expression convoluted with behavioral and cultural 'languages' requiring a significant extension of the individual-oriented formal theory of the earlier chapters. This would, perhaps, be necessary in any case, since developing multicellular organisms must inevitably undergo collective processes of gene expression.

12.5 Acknowledgment

This work was partially supported by an NIH grant to CUES: The Neighborhood Environment and Drug Use in New York City, RO1 DA 017642.

12.6 Chapter Appendix

Data Sets Used and their Sources

Health-Related Data Sets
Deaths from selected causes in 2000
Number of deaths per 100,000 in each community district:
Cardiovascular and other diseases of the heart
Cancer
HIV/AIDS
Influenza/pneumonia
Stroke
Chronic lower respiratory diseases
Chronic liver disease and cirrhosis
Diabetes
Drugs
Accidents
Suicides
Homicides
Births and their characteristics:
Infant mortality rate (per 1000 live births)
Percent low-weight births
Percent live births to teenagers
Percent live births to single mothers
Source: Summary of Vital Statistics 2000: The City of New York Office of Vital Statistics, New York City Health Department
Neighborhood Environment:
Percent of streets rated acceptably clean
Number of structural fires
Source: downloaded from Mayors Office of Operations Neighborhood Statistics Neighborhood Housing
Total number of occupied housing units, 1999
Number of occupied housing units with 3 or more serious violations, 1999
Source: NYC Housing and Vacancy Survey
Demographics and Socioeconomic Measures:
Total population
Population of following ethnicities: white, black, Hispanic, Asian, native American Pacific Islander, other, mixed
Population of the following age ranges: under 5, 5-9, 10-14, 15-19, 20-24, 25-34, 35-34, 45-54, 55-59, 60-64, 65-74, 75-84, 85 and older
Population over 25 with the following educational attainments: less than 9th grade, Some high school (no diploma), high school, some college (no degree) Associate degree, bachelors degree, and graduate degree.
Unemployed persons over 16 years of age
Total number of households
Number of households with the following income ranges: <$10k, $10,000-14,999, $15,000-$24,999, $25,000-$34,999, $35,000-$49,999, $50,000-$74,999, $75,000-$99,999, $100,000-$149,999, $150,000-$199,999, $200k or more.
Median household income in 1999

Source: Data on total population and population by ethnicity and age came from the Department of City Planning 2000 Census Community District Demographic Tables. Data on unemployment, educational attainment, household income, and median household income came from the Department of City Planning 2000 Census Community District Socioeconomic Tables.

13
Final thoughts

We have presented an epigenetic catalysis model of development that can, in a natural manner, account for individual or simple group aggregate scale phenomena. The ideas fit well to observed data on the US obesity pandemic, its correlates of coronary heart disease, hypertension, and diabetes, and possibly to some cancers and autoimmune disorders. The model fits less well at the neighborhood scale, where multiple chronic diseases not only emerge, but may interact and reflect phenomena of group gene expression. These may possibly be analogous to the mesoscale resonance effects of Wallace and Wallace (2008), driven by the interaction of the three types of human heritage: culture, genetic, and epigenetic. The Upper Manhattan study, that used a carefully selected sample of high function young mothers as indicators of neighborhood conditions, required some fairly subtle argument to explain figure 11.2. Chapter 12 found a neighborhood level disease guild of chronic and behavioral disorders that appears to display emergent properties among interacting processes of behavior and gene expression beyond the ability of our model to easily explain.

A developing multicellular organism would, perhaps, show analogous patterns of collective gene expression. For humans, individual and group dynamics are further affected by the complication of embedding 'languages' of culture. We have, then, driven a cutting-edge theoretical paradigm to its limit, and indicated a direction for generalization, involving an extension toward understanding cultural 'lock-in' of simpler epigenetic mechanisms, and this is no small thing.

What is clear, certainly, is that epigenetic catalysis can serve as a unifying theme across a plethora of gene expression phenomena. For human populations, psychosocial stressors are highly plieotropic catalytic forces writing images of themselves as a broad spectrum of chronic diseases. The synergism of deindustrialization and population displacement – economic and geographic disruption – seems adequate to explain much of the 'mystery' of the US obesity epidemic and its many health sequelae. Given the nature of the collective catalytic processes, it seems unlikely that individual-oriented medical or behavioral interventions will be able to ameliorate these impacts. As in the past, these will, however, serve the political purpose of diverting attention from large scale structural dynamics driving the deterioration in public health, and basically blame the victims of those dynamics as 'not taking responsibility' for their health and behavior. Blaming the victim is, we all know, as American as apple pie and violence. The processes producing figures 7.1 and 7.3 represent the failings of public policy and economic structures, not the failings of individuals entrapped by those policies and structures.

More specifically, the plieotropy inherent to epigenetic catalysis by psychosocial stress implies that even the best-intentioned individual-oriented, or large scale individual disease-oriented, interventions risk becoming little more than politically motivated whack-a-mole games, giving stressed populations the 'choice' of early death by different chronic diseases. This is unlikely to change in the absence of large scale political and social reform in the United States, and, as figures 7.7 and 7.8 show, will enmesh dominant as well as subordinate populations in an overall and closely interlinked dynamic of public health collapse. Those figures illustrate the 'paradox of Apartheid' in that all sectors of the system of American Apartheid (sensu Massey and Denton, 1992) inevitably suffer, in spite of the powerful cultural illusion that Apartheid protects 'us' from 'them'.

The mathematical treatment of epigenetic catalysis could serve as a foundation for the formal programming of artificial coevolutionary systems in biotechnology. The epigenetic generalized enzyme is an information source analogous to a program in a computing machine, serving to channel the output of a highly dynamic artificial biosystem. Elsewhere, we have suggested a similar approach to programming highly parallel coevolutionary computing devices (Wallace, 2009).

From another perspective, we have introduced a new class of probability models of organismal development and its dysfunctions that can be converted to statistical tools for data analysis. This is difficult, as those who work to manufacture more conventional statistical machinery will attest, and seems, along with understanding cultural lock-in – the effect of the third system of heritage in human populations – a most compelling next step for the mathematical workplan.

Although Glazebrook and Wallace (2009) present more mathematical detail, we hope this book can serve as an introduction, convincing the reader that these and related methods might be worth pursuing further.

14
Mathematical appendix

14.1 The Shannon Coding Theorem

Messages from a source, seen as symbols x_j from some alphabet, each having probabilities P_j associated with a random variable X, are 'encoded' into the language of a 'transmission channel', a random variable Y with symbols y_k, having probabilities P_k, possibly with error. Someone receiving the symbol y_k then retranslates it (without error) into some x_k, which may or may not be the same as the x_j that was sent.

More formally, the message sent along the channel is characterized by a random variable X having the distribution

$$P(X = x_j) = P_j, j = 1, ..., M.$$

The channel through which the message is sent is characterized by a second random variable Y having the distribution

$$P(Y = y_k) = P_k, k = 1, ..., L.$$

Let the joint probability distribution of X and Y be defined as

$$P(X = x_j, Y = y_k) = P(x_j, y_k) = P_{j,k}$$

and the conditional probability of Y given X as

$$P(Y = y_k | X = x_j) = P(y_k | x_j).$$

Then the Shannon uncertainty of X and Y independently and the joint uncertainty of X and Y together are defined respectively as

$$H(X) = -\sum_{j=1}^{M} P_j \log(P_j)$$

$$H(Y) = -\sum_{k=1}^{L} P_k \log(P_k)$$

$$H(X,Y) = -\sum_{j=1}^{M}\sum_{k=1}^{L} P_{j,k} \log(P_{j,k}).$$

(14.1)

The *conditional uncertainty* of Y given X is defined as

$$H(Y|X) = -\sum_{j=1}^{M}\sum_{k=1}^{L} P_{j,k} \log[P(y_k|x_j)]$$

(14.2)

For any two stochastic variates X and Y, $H(Y) \geq H(Y|X)$, as knowledge of X generally gives some knowledge of Y. Equality occurs only in the case of stochastic independence.

Since $P(x_j, y_k) = P(x_j)P(y_k|x_j)$, we have

$$H(X|Y) = H(X,Y) - H(Y)$$

The information transmitted by translating the variable X into the channel transmission variable Y – possibly with error – and then retranslating without error the transmitted Y back into X is defined as

$$I(X|Y) \equiv H(X) - H(X|Y) = H(X) + H(Y) - H(X,Y)$$

(14.3)

See, for example, Ash (1990), Khinchin (1957) or Cover and Thomas (1991) for details. The essential point is that if there is no uncertainty in X given the channel Y, then there is no loss of information through transmission.

In general this will not be true, and herein lies the essence of the theory.

Given a fixed vocabulary for the transmitted variable X, and a fixed vocabulary and probability distribution for the channel Y, we may vary the probability distribution of X in such a way as to maximize the information sent. The capacity of the channel is defined as

$$C \equiv \max_{P(X)} I(X|Y)$$

(14.4)

subject to the subsidiary condition that $\sum P(X) = 1$.

The critical trick of the Shannon Coding Theorem for sending a message with arbitrarily small error along the channel Y at any rate $R < C$ is to encode it in longer and longer 'typical' sequences of the variable X; that is, those sequences whose distribution of symbols approximates the probability distribution $P(X)$ above which maximizes C.

If $S(n)$ is the number of such 'typical' sequences of length n, then

$$\log[S(n)] \approx nH(X)$$

where $H(X)$ is the uncertainty of the stochastic variable defined above. Some consideration shows that $S(n)$ is much less than the total number of possible messages of length n. Thus, as $n \to \infty$, only a vanishingly small fraction of all possible messages is meaningful in this sense. This observation, after some considerable development, is what allows the Coding Theorem to work so well. In sum, the prescription is to encode messages in typical sequences, which are sent at very nearly the capacity of the channel. As the encoded messages become longer and longer, their maximum possible rate of transmission without error approaches channel capacity as a limit. Again, Ash (1990), Khinchin (1957) and Cover and Thomas (1991) provide details.

14.2 The 'tuning theorem'

Telephone lines, optical wave guides and the tenuous plasma through which a planetary probe transmits data to earth may all be viewed in traditional information-theoretic terms as a *noisy channel* around which we must structure a message so as to attain an optimal error-free transmission rate.

Telephone lines, wave guides and interplanetary plasmas are, relatively speaking, fixed on the timescale of most messages, as are most sociogeographic networks. Indeed, the capacity of a channel, according to equation 14.4, is defined by varying the probability distribution of the 'message' process X so as to maximize $I(X|Y)$.

Suppose there is some message X so critical that its probability distribution must remain fixed. The trick is to fix the distribution $P(x)$ but *modify the channel* – i.e. tune it – so as to maximize $I(X|Y)$. The *dual* channel capacity C^* can be defined as

$$C^* \equiv \max_{P(Y),P(Y|X)} I(X|Y)$$

(14.5)

But

$$C^* = \max_{P(Y),P(Y|X)} I(Y|X)$$

since

$$I(X|Y) = H(X) + H(Y) - H(X,Y) = I(Y|X).$$

Thus, in a purely formal mathematical sense, *the message transmits the channel*, and there will indeed be, according to the Coding Theorem, a channel distribution $P(Y)$ which maximizes C^*.

One may do better than this, however, by modifying the channel matrix $P(Y|X)$. Since

$$P(y_j) = \sum_{i=1}^{M} P(x_i) P(y_j|x_i),$$

$P(Y)$ is entirely defined by the channel matrix $P(Y|X)$ for fixed $P(X)$ and

$$C^* = \max_{P(Y),P(Y|X)} I(Y|X) = \max_{P(Y|X)} I(Y|X).$$

Calculating C^* requires maximizing the complicated expression

$$I(X|Y) = H(X) + H(Y) - H(X,Y)$$

which contains products of terms and their logs, subject to constraints that the sums of probabilities are 1 and each probability is itself between 0 and 1. Maximization is done by varying the channel matrix terms $P(y_j|x_i)$ within the constraints. This is a difficult problem in nonlinear optimization requiring Lagrange multiplier methods. However, for the special case $M = L$, C^* may be found by inspection:

If $M = L$, then choose

$$P(y_j|x_i) = \delta_{j,i}$$

where $\delta_{i,j}$ is 1 if $i = j$ and 0 otherwise. For this special case

$$C^* \equiv H(X)$$

with $P(y_k) = P(x_k)$ for all k. *Information is thus transmitted without error when the channel becomes 'typical' with respect to the fixed message distribution $P(X)$.*

If $M < L$ matters reduce to this case, but for $L < M$ information must be lost, leading to Rate Distortion arguments explored more fully below.

Thus modifying the channel may be a far more efficient means of ensuring transmission of an important message than encoding that message in a 'natural' language which maximizes the rate of transmission of information on a fixed channel.

We have examined the two limits in which either the distributions of $P(Y)$ or of $P(X)$ are kept fixed. The first provides the usual Shannon Coding Theorem, and the second, hopefully, a tuning theorem variant. It seems likely, however, than for many important systems $P(X)$ and $P(Y)$ will 'interpenetrate,' to use Richard Levins' terminology. That is, $P(X)$ and $P(Y)$ will affect each other in characteristic ways, so that some form of mutual tuning may be the most effective strategy.

14.3 The Shannon-McMillan Theorem

Not all statements – sequences of the random variable X – are equivalent. According to the structure of the underlying language of which the message is a particular expression, some messages are more 'meaningful' than others, that is, in accord with the grammar and syntax of the language. The other principal result from information theory, the Shannon-McMillan or Asymptotic Equipartition Theorem, describes how messages themselves are to be classified.

Suppose a long sequence of symbols is chosen, using the output of the random variable X above, so that an output sequence of length n, with the form

$$x_n = (\alpha_0, \alpha_1, ..., \alpha_{n-1})$$

has joint and conditional probabilities

$$P(X_0 = \alpha_0, X_1 = \alpha_1, ..., X_{n-1} = \alpha_{n-1})$$

$$P(X_n = \alpha_n | X_0 = \alpha_0, ..., X_{n-1} = \alpha_{n-1}).$$

(14.6)

Using these probabilities we may calculate the conditional uncertainty

$$H(X_n | X_0, X_1, ..., X_{n-1}).$$

The uncertainty of the *information source*, $H[\mathbf{X}]$, is defined as

$$H[\mathbf{X}] \equiv \lim_{n \to \infty} H(X_n | X_0, X_1, ..., X_{n-1}).$$

(14.7)

In general

$$H(X_n | X_0, X_1, ..., X_{n-1}) \leq H(X_n).$$

Only if the random variables X_j are all stochastically independent does equality hold. If there is a maximum n such that, for all $m > 0$

$$H(X_{n+m} | X_0, ..., X_{n+m-1}) = H(X_n | X_0, ..., X_{n-1}),$$

then the source is said to be of *order* n. It is easy to show that

$$H[\mathbf{X}] = \lim_{n \to \infty} \frac{H(X_0, ...X_n)}{n+1}.$$

In general the outputs of the $X_j, j = 0, 1, ..., n$ are *dependent*. That is, the output of the communication process at step n depends on previous steps. Such serial correlation, in fact, is the very structure which enables most of what follows in this book.

Here, however, the processes are all assumed fixed, that is, the serial correlations do not change in time, and the system is *stationary*.

14.3 The Shannon-McMillan Theorem

A very broad class of such self-correlated, stationary, information sources, the so-called *ergodic* sources for which the long-run relative frequency of a sequence converges stochastically to the probability assigned to it, have a particularly interesting property:

It is possible, in the limit of large n, to divide all sequences of outputs of an ergodic information source into two distinct sets, S_1 and S_2, having, respectively, very high and very low probabilities of occurrence, with the source uncertainty providing the splitting criterion. In particular the Shannon-McMillan Theorem states that, for a (long) sequence having n (serially correlated) elements, the number of 'meaningful' sequences, $N(n)$ – those belonging to set S_1 – will satisfy the relation

$$\frac{\log[N(n)]}{n} \approx H[\mathbf{X}].$$

(14.8)

More formally,

$$\lim_{n \to \infty} \frac{\log[N(n)]}{n} = H[\mathbf{X}]$$

$$= \lim_{n \to \infty} H(X_n | X_0, ..., X_{n-1})$$

$$= \lim_{n \to \infty} \frac{H(X_0, ..., X_n)}{n+1}.$$

(14.9)

The Shannon Coding theorem, by means of an analogous splitting argument, shows that for any rate $R < C$, where C is the channel capacity, a message may be sent without error, using the probability distribution for X which maximizes $I(X|Y)$ as the coding scheme. Using the internal structures of the information source permits *limiting attention only to meaningful sequences of symbols*. This restriction can greatly raise the maximum possible rate at which information can be transmitted with arbitrarily small error: if there are M possible symbols and the uncertainty of the source is $H[\mathbf{X}]$, then the effective capacity of the channel C, using this 'source coding,' becomes (Ash, 1990)

$$C_E = C \frac{\log(M)}{H[\mathbf{X}]}.$$

(14.10)

As $H[\mathbf{X}] \leq \log(M)$, with equality only for stochastically independent, uniformly distributed random variables,

$$C_E \geq C.$$

(14.11)

Note that, for a given channel capacity, the condition

$$H[\mathbf{X}] \leq C$$

always holds.

14.4 The Rate Distortion Theorem

The Shannon-McMillan Theorem can be expressed as the 'zero error limit' of the Rate Distortion Theorem (Dembo and Zeitouni, 1998; Cover and Thomas, 1991). The theorem defines a splitting criterion that identifies high probability pairs of sequences. We follow closely the treatment of Cover and Thomas (1991).

The origin of the problem is the question of representing one information source by a simpler one in such a way that the least information is lost. For example we might have a continuous variate between 0 and 100, and wish to represent it in terms of a small set of integers in a way that minimizes the inevitable distortion that process creates. Typically, for example, an analog audio signal will be replaced by a 'digital' one. The problem is to do this in a way which least distorts the *reconstructed* audio waveform.

Suppose the original stationary, ergodic information source Y with output from a particular alphabet generates sequences of the form

14.4 The Rate Distortion Theorem

$$y^n = y_1, ..., y_n.$$

These are 'digitized,' in some sense, producing a chain of 'digitized values'

$$b^n = b_1, ..., b_n,$$

where the b-alphabet is much more restricted than the y-alphabet.

b^n is, in turn, *deterministically retranslated* into a reproduction of the original signal y^n. That is, each b^m is mapped on to a unique n-length y-sequence in the alphabet of the information source Y:

$$b^m \to \hat{y}^n = \hat{y}_1, ..., \hat{y}_n.$$

Note, however, that many y^n sequences may be mapped onto the *same* retranslation sequence $\hat{y}^n$, so that information will, in general, be lost.

The central problem is to explicitly minimize that loss.

The retranslation process defines a new stationary, ergodic information source, $\hat{Y}$.

The next step is to define a *distortion measure*, $d(y, \hat{y})$, which compares the original to the retranslated path. For example the *Hamming distortion* is

$$d(y, \hat{y}) = 1, y \neq \hat{y}$$

$$d(y, \hat{y}) = 0, y = \hat{y}.$$

(14.12)

For continuous variates the *Squared error distortion* is

$$d(y, \hat{y}) = (y - \hat{y})^2.$$

(14.13)

Possibilities abound.

The distortion between paths y^n and $\hat{y}^n$ is defined as

$$d(y^n,\hat{y}^n) = \frac{1}{n}\sum_{j=1}^{n} d(y_j,\hat{y}_j).$$

(14.14)

Suppose that with each path y^n and b^n-path retranslation into the y-language and denoted y^n, there are associated individual, joint, and conditional probability distributions

$$p(y^n), p(\hat{y}^n), p(y^n|\hat{y}^n).$$

The *average distortion* is defined as

$$D = \sum_{y^n} p(y^n) d(y^n,\hat{y}^n).$$

(14.15)

It is possible, using the distributions given above, to define the information transmitted from the incoming Y to the outgoing $\hat{Y}$ process in the usual manner, using the Shannon source uncertainty of the strings:

$$I(Y,\hat{Y}) \equiv H(Y) - H(Y|\hat{Y}) = H(Y) + H(\hat{Y}) - H(Y,\hat{Y}).$$

If there is no uncertainty in Y given the retranslation $\hat{Y}$, then no information is lost.

In general, this will not be true.

The *information rate distortion function* $R(D)$ for a source Y with a distortion measure $d(y,\hat{y})$ is defined as

$$R(D) = \min_{p(y,\hat{y});\sum_{(y,\hat{y})} p(y)p(y|\hat{y})d(y,\hat{y}) \leq D} I(Y,\hat{Y}).$$

(14.16)

The minimization is over all conditional distributions $p(y|\hat{y})$ for which the joint distribution $p(y,\hat{y}) = p(y)p(y|\hat{y})$ satisfies the average distortion constraint (i.e. average distortion $\leq D$).

The *Rate Distortion Theorem* states that $R(D)$ *is the maximum achievable rate of information transmission which does not exceed the distortion D*. Cover and Thomas (1991) or Dembo and Zeitouni (1998) provide details.

An important observation is that $R(D)$ is necessarily convex in D (Cover and Thomas, 1991, Lemma 13.4.1). This has profound implications for dynamic processes, since R can be interpreted as a free energy homolog, as it is a channel capacity measure.

It is also important to note that pairs of sequences $(y^n, \hat{y}^n)$ can be defined as *distortion typical*; that is, for a given average distortion D, defined in terms of a particular measure, pairs of sequences can be divided into two sets, a high probability one containing a relatively small number of (matched) pairs with $d(y^n, \hat{y}^n) \leq D$, and a low probability one containing most pairs. As $n \to \infty$, the smaller set approaches unit probability, and, for those pairs,

$$p(y^n) \geq p(\hat{y}^n | y^n) \exp[-nI(Y, \hat{Y})].$$

(14.17)

Thus, roughly speaking, $I(Y, \hat{Y})$ embodies the splitting criterion between high and low probability pairs of paths.

For the theory of interacting information sources, then, $I(Y, \hat{Y})$ can play the role of H in a generalized Onsager relations argumen t.

The rate distortion function can actually be calculated in many cases by using a Lagrange multiplier method – see Section 13.7 of Cover and Thomas (1991).

Glazebrook and Wallace (2009) suggest using something like $s \equiv d(\hat{x}, x)$ as a metric in a geometry of information sources, e.g. when simple ergodicity fails, and $H(x) \neq H(\hat{x})$ for high probability paths $\hat{x}$ and x.

14.5 Groupoids

Basic ideas

Following Weinstein (1996) closely, a groupoid, G, is defined by a base set A upon which some mapping – a morphism – can be defined. Note that not all possible pairs of states (a_j, a_k) in the base set A can be connected by such a morphism. Those that can define the groupoid element, a morphism $g = (a_j, a_k)$ having the natural inverse $g^{-1} = (a_k, a_j)$. Given such a pairing, it is possible to define 'natural' end-point maps $\alpha(g) = a_j, \beta(g) = a_k$ from the set of morphisms G into A, and a formally associative product in the groupoid $g_1 g_2$ provided $\alpha(g_1 g_2) = \alpha(g_1), \beta(g_1 g_2) = \beta(g_2)$, and $\beta(g_1) = \alpha(g_2)$. Then the product is defined, and associative, $(g_1 g_2) g_3 = g_1 (g_2 g_3)$.

In addition, there are natural left and right identity elements λ_g, ρ_g such that $\lambda_g g = g = g\rho_g$.

An orbit of the groupoid G over A is an equivalence class for the relation $a_j \sim Ga_k$ if and only if there is a groupoid element g with $\alpha(g) = a_j$ and $\beta(g) = a_k$. Following Cannas DaSilva and Weinstein (1999),, we note that a groupoid is called transitive if it has just one orbit. The transitive groupoids are the building blocks of groupoids in that there is a natural decomposition of the base space of a general groupoid into orbits. Over each orbit there is a transitive groupoid, and the disjoint union of these transitive groupoids is the original groupoid. Conversely, the disjoint union of groupoids is itself a groupoid.

The isotropy group of $a \in X$ consists of those g in G with $\alpha(g) = a = \beta(g)$. These groups prove fundamental to classifying groupoids.

If G is any groupoid over A, the map $(\alpha, \beta) : G \to A \times A$ is a morphism from G to the pair groupoid of A. The image of (α, β) is the orbit equivalence relation $\sim G$, and the functional kernel is the union of the isotropy groups. If $f : X \to Y$ is a function, then the kernel of f, $ker(f) = [(x_1, x_2) \in X \times X : f(x_1) = f(x_2)]$ defines an equivalence relation.

Groupoids may have additional structure. As Weinstein (1996) explains, a groupoid G is a topological groupoid over a base space X if G and X are topological spaces and α, β and multiplication are continuous maps. A criticism sometimes applied to groupoid theory is that their classification up to isomorphism is nothing other than the classification of equivalence relations via the orbit equivalence relation and groups via the isotropy groups. The imposition of a compatible topological structure produces a nontrivial interaction between the two structures. It is possible to introduce a metric structure on manifolds of related information sources, producing such interaction.

In essence, a groupoid is a category in which all morphisms have an inverse, here defined in terms of connection to a base point by a meaningful path of an information source dual to a cognitive process.

As Weinstein points out, the morphism (α, β) suggests another way of looking at groupoids. A groupoid over A identifies not only which elements of A are equivalent to one another (isomorphic), but *it also parametrizes the different ways (isomorphisms) in which two elements can be equivalent*, i.e., all possible information sources dual to some cognitive process. Given the information theoretic characterization of cognition presented above, this produces a full modular cognitive network in a highly natural manner.

Brown (1987) describes the fundamental structure as follows:

> A groupoid should be thought of as a group with many objects, or with many identities... A groupoid with one object is essentially just a group. So the notion of groupoid is an extension of that of groups. It gives an additional convenience, flexibility and range of applications...
> EXAMPLE 1. A disjoint union [of groups] $G = \cup_\lambda G_\lambda, \lambda \in \Lambda$, is a groupoid: the product ab is defined if and only if a, b belong to the

same G_λ, and ab is then just the product in the group G_λ. There is an identity 1_λ for each $\lambda \in \Lambda$. The maps α, β coincide and map G_λ to λ, $\lambda \in \Lambda$.

EXAMPLE 2. An equivalence relation R on [a set] X becomes a groupoid with $\alpha, \beta : R \to X$ the two projections, and product $(x,y)(y,z) = (x,z)$ whenever $(x,y), (y,z) \in R$. There is an identity, namely (x,x), for each $x \in X$...

Weinstein makes the following fundamental point:

> Almost every interesting equivalence relation on a space B arises in a natural way as the orbit equivalence relation of some groupoid G over B. Instead of dealing directly with the orbit space B/G as an object in the category S_{map} of sets and mappings, one should consider instead the groupoid G itself as an object in the category G_{htp} of groupoids and homotopy classes of morphisms.

The groupoid approach has become quite popular in the study of networks of coupled dynamical systems which can be defined by differential equation models, (e.g., Golubitsky and Stewart, 2006).

Global and local symmetry groupoids

Again we follow Weinstein (1996) fairly closely, using his example of a finite tiling.

Consider a tiling of the euclidean plane R^2 by identical 2 by 1 rectangles, specified by the set X (one dimensional) where the grout between tiles is $X = H \cup V$, having $H = R \times Z$ and $V = 2Z \times R$, where R is the set of real numbers and Z the integers. Call each connected component of $R^2 \backslash X$, that is, the complement of the two dimensional real plane intersecting X, a tile.

Let Γ be the group of those rigid motions of R^2 which leave X invariant, i.e., the normal subgroup of translations by elements of the lattice $\Lambda = H \cap V = 2Z \times Z$ (corresponding to corner points of the tiles), together with reflections through each of the points $1/2\Lambda = Z \times 1/2Z$, and across the horizontal and vertical lines through those points. As noted in [78], much is lost in this coarse-graining, in particular the same symmetry group would arise if we replaced X entirely by the lattice Λ of corner points. Γ retains no information about the local structure of the tiled plane. In the case of a real tiling, restricted to the finite set $B = [0, 2m] \times [0, n]$ the symmetry group shrinks drastically: The subgroup leaving $X \cap B$ invariant contains just four elements even though a repetitive pattern is clearly visible. A two-stage groupoid approach recovers the lost structure.

We define the transformation groupoid of the action of Γ on R^2 to be the set

$$G(\Gamma, R^2) = \{(x, \gamma, y | x \in R^2, y \in R^2, \gamma \in \Gamma, x = \gamma y\},$$

with the partially defined binary operation

$$(x, \gamma, y)(y, \nu, z) = (x, \gamma\nu, z).$$

Here $\alpha(x, \gamma, y) = x$, and $\beta(x, \gamma, y) = y$, and the inverses are natural.

We can form the restriction of G to B (or any other subset of R^2) by defining

$$G(\Gamma, R^2)|_B = \{g \in G(\Gamma, R^2) | \alpha(g), \beta(g) \in B\}$$

[1]. An orbit of the groupoid G over B is an equivalence class for the relation

$x \sim_G y$ if and only if there is a groupoid element g with $\alpha(g) = x$ and $\beta(g) = y$.

Two points are in the same orbit if they are similarly placed within their tiles or within the grout pattern.

[2]. The isotropy group of $x \in B$ consists of those g in G with $\alpha(g) = x = \beta(g)$. It is trivial for every point except those in $1/2\Lambda \cap B$, for which it is $Z_2 \times Z_2$, the direct product of integers modulo two with itself.

By contrast, embedding the tiled structure within a larger context permits definition of a much richer structure, i.e., the identification of local symmetries.

We construct a second groupoid as follows. Consider the plane R^2 as being decomposed as the disjoint union of $P_1 = B \cap X$ (the grout), $P_2 = B \backslash P_1$ (the complement of P_1 in B, which is the tiles), and $P_3 = R^2 \backslash B$ (the exterior of the tiled room). Let E be the group of all euclidean motions of the plane, and define the local symmetry groupoid G_{loc} as the set of triples (x, γ, y) in $B \times E \times B$ for which $x = \gamma y$, and for which y has a neighborhood $\mathcal{U}$ in R^2 such that $\gamma(\mathcal{U} \cap P_i) \subseteq P_i$ for $i = 1, 2, 3$. The composition is given by the same formula as for $G(\Gamma, R^2)$.

For this groupoid-in-context there are only a finite number of orbits:

$\mathcal{O}_1$ = interior points of the tiles.
$\mathcal{O}_2$ = interior edges of the tiles.
$\mathcal{O}_3$ = interior crossing points of the grout.
$\mathcal{O}_4$ = exterior boundary edge points of the tile grout.
$\mathcal{O}_5$ = boundary 'T' points.
$\mathcal{O}_6$ = boundary corner points.

The isotropy group structure is, however, now very rich indeed:

The isotropy group of a point in $\mathcal{O}_1$ is now isomorphic to the entire rotation group O_2.

It is $Z_2 \times Z_2$ for $\mathcal{O}_2$.

For $\mathcal{O}_3$ it is the eight-element dihedral group D_4.

For $\mathcal{O}_4, \mathcal{O}_5$ and $\mathcal{O}_6$ it is simply Z_2.

These are the 'local symmetries' of the tile-in-context.

14.6 Morse Theory

Morse theory examines relations between analytic behavior of a function – the location and character of its critical points – and the underlying topology of the manifold on which the function is defined. We are interested in a number of such functions, for example information source uncertainty on a parameter space and 'second order' iterations involving parameter manifolds determining critical behavior, for example sudden onset of a giant component in the mean number model (Wallace and Wallace, 2008), and universality class tuning in a mean field model. These can be reformulated from a Morse theory perspective. Here we follow closely the elegant treatments of Pettini (2007) and Kastner (2006).

The essential idea of Morse theory is to examine an n-dimensional manifold M as decomposed into level sets of some function $f : M \to \mathbf{R}$ where $\mathbf{R}$ is the set of real numbers. The a-level set of f is defined as

$$f^{-1}(a) = \{x \in M : f(x) = a\},$$

the set of all points in M with $f(x) = a$. If M is compact, then the whole manifold can be decomposed into such slices in a canonical fashion between two limits, defined by the minimum and maximum of f on M. Let the part of M below a be defined as

$$M_a = f^{-1}(-\infty, a] = \{x \in M : f(x) \le a\}.$$

These sets describe the whole manifold as a varies between the minimum and maximum of f.

Morse functions are defined as a particular set of smooth functions $f : M \to \mathbf{R}$ as follows. Suppose a function f has a critical point x_c, so that the derivative $df(x_c) = 0$, with critical value $f(x_c)$. Then f is a Morse function if its critical points are nondegenerate in the sense that the Hessian matrix of second derivatives at x_c, whose elements, in terms of local coordinates are

$$\mathcal{H}_{i,j} = \partial^2 f / \partial x^i \partial x^j,$$

has rank n, which means that it has only nonzero eigenvalues, so that there are no lines or surfaces of critical points and, ultimately, critical points are isolated.

The index of the critical point is the number of negative eigenvalues of $\mathcal{H}$ at x_c.

A level set $f^{-1}(a)$ of f is called a critical level if a is a critical value of f, that is, if there is at least one critical point $x_c \in f^{-1}(a)$.

Again following Pettini (2007), the essential results of Morse theory are:

[1] If an interval $[a, b]$ contains no critical values of f, then the topology of $f^{-1}[a, v]$ does not change for any $v \in (a, b]$. Importantly, the result is valid even if f is not a Morse function, but only a smooth function.

[2] If the interval $[a, b]$ contains critical values, the topology of $f^{-1}[a, v]$ changes in a manner determined by the properties of the matrix H at the critical points.

[3] If $f : M \to \mathbf{R}$ is a Morse function, the set of all the critical points of f is a discrete subset of M, i.e., critical points are isolated. This is Sard's Theorem.

[4] If $f : M \to \mathbf{R}$ is a Morse function, with M compact, then on a finite interval $[a,b] \subset \mathbf{R}$, there is only a finite number of critical points p of f such that $f(p) \in [a,b]$. The set of critical values of f is a discrete set of $\mathbf{R}$.

[5] For any differentiable manifold M, the set of Morse functions on M is an open dense set in the set of real functions of M of differentiability class r for $0 \leq r \leq \infty$.

[6] Some topological invariants of M, that is, quantities that are the same for all the manifolds that have the same topology as M, can be estimated and sometimes computed exactly once all the critical points of f are known: Let the Morse numbers $\mu_i (i = 0, ..., m)$ of a function f on M be the number of critical points of f of index i, (the number of negative eigenvalues of H). The Euler characteristic of the complicated manifold M can be expressed as the alternating sum of the Morse numbers of any Morse function on M,

$$\chi = \sum_{i=1}^{m} (-1)^i \mu_i.$$

The Euler characteristic reduces, in the case of a simple polyhedron, to

$$\chi = V - E + F$$

where $V, E,$ and F are the numbers of vertices, edges, and faces in the polyhedron.

[7] Another important theorem states that, if the interval $[a,b]$ contains a critical value of f with a single critical point x_c, then the topology of the set M_b defined above differs from that of M_a in a way which is determined by the index, i, of the critical point. Then M_b is homeomorphic to the manifold obtained from attaching to M_a an i-handle, i.e., the direct product of an i-disk and an $(m-i)$-disk.

Again, see Pettini (2007) or Matsumoto (2002) for details.

14.7 Generalized Onsager Theory

Understanding the time dynamics of groupoid-driven information systems away from phase transition critical points requires a phenomenology similar to the Onsager relations of nonequilibrium thermodynamics. This also leads to a general theory involving large-scale topological changes in the sense of Morse theory.

If the Groupoid Free Energy (GFE) of a biological process is parametized by some vector of quantities $\mathbf{K} \equiv (K_1, ..., K_m)$, then, in analogy with nonequilibrium thermodynamics, gradients in the K_j of the *disorder*, defined as

$$S_G \equiv F_G(\mathbf{K}) - \sum_{j=1}^{m} K_j \partial F_G/\partial K_j$$

(14.18)

become of central interest.

Equation 14.18 is similar to the definition of entropy in terms of the free energy of a physical system.

Pursuing the homology further, the generalized Onsager relations defining temporal dynamics of systems having a GFE become

$$dK_j/dt = \sum_i L_{j,i} \partial S_G/\partial K_i,$$

(14.19)

where the $L_{j,i}$ are, in first order, constants reflecting the nature of the underlying cognitive phenomena. The L-matrix is to be viewed empirically, in the same spirit as the slope and intercept of a regression model, and may have structure far different than familiar from more simple chemical or physical processes. The $\partial S_G/\partial K$ are analogous to thermodynamic forces in a chemical system, and may be subject to override by external physiological or other driving mechanisms: biological and cognitive phenomena, unlike simple physical systems, can make choices as to resource allocation.

That is, an essential contrast with simple physical systems driven by (say) entropy maximization is that complex biological or cognitive structures can make decisions about resource allocation, to the extent resources are available. Thus resource availability is a context, not a determinant, of behavior.

Equations 14.18 and 14.19 can be derived in a simple parameter-free covariant manner which relies on the underlying topology of the information source space implicit to the development Wallace and Wallace (2008). We will not pursue that development here. See Glazebrook and Wallace (2009) for details.

The dynamics, as we have presented them so far, have been noiseless, while biological systems are always very noisy. Equation 14.19 might be rewritten as

$$dK_j/dt = \sum_i L_{j,i}\partial S_G/\partial K_i + \sigma W(t)$$

where σ is a constant and $W(t)$ represents white noise. This leads directly to a family of classic stochastic differential equations having the form

$$dK_t^j = L^j(t,\mathbf{K})dt + \sigma^j(t,\mathbf{K})dB_t,$$

(14.20)

where the L^j and σ^j are appropriately regular functions of t and $\mathbf{K}$, and dB_t represents the noise structure, and we have readjusted the indices.

Further progress in this direction requires introduction of methods from stochastic differential geometry and related topics in the sense of Emery (1989). The obvious inference is that noise – not necessarily 'white' – can serve as a tool to shift the system between various topological modes, as a kind of crosstalk and the source of a generalized stochastic resonance.

Effectively, topological shifts between and within dynamic manifolds constitute another theory of phase transitions (Pettini, 2007), and this phenomenological Onsager treatment would likely be much enriched by explicit adoption of a Morse theory perspective.

15
References

Abler, R., Adams, J., Gould, P.: Spatial Organization: The Geographer's View of the World. Prentice Hall: Englewood Cliffs (1971).

Adami, C., Cerf, N.: Physical complexity of symbolic sequences, Physica D, 137:62-69 (2000).

Adami, C., Ofria, C., Collier, T.: Evolution of biological complexity, Proceedings of the National Academy of Sciences, 97:4463-4468 (2000).

Ahlborg, A., Ljung, T., Rosmond, R., McEwen, B., Holm, G., Akesson, H., Bjorntorp P.: Depression and Anxiety symptoms in relation to anthropometry and metabolism in men. Psychiatry Research, 112:101-110 (2002).

Albert R., Barabasi, A.: Statistical mechanics of complex networks, Reviews of Modern Physics, 74:47-97 (2002).

Allan, J.: Explanatory models of overweight among African American, Euro-American, and Mexican American women, Western Journal of Nursing Research, 20:45-66, (1998).

Andersen, H.R. et al. (27 other authors): Comparison of short-term Estrogenicity tests for identification of hormone-disrupting chemicals, Environmental Health Perspectives, 107 (supp): 89-108 (1999).

Annesi-Maesan, I., Moreau, D., Strachan, D.: In utero and perinatal complications preceding asthma. Allergy, 56:491-497 (2001).

Antonijevic, I., Murck, H., Frieboes, R., Horn, R., Brabant, G., Steiger, A.: Elevated nocturnal profiles Of serum leptin in patients with depression. Psychiatric Research, 32:403-410 (1998).

Ash R.: Information Theory. Dover Publications, New York (1990).

Atlan, H., Cohen, I.: Immune information, self-organization, and meaning. International Immunology 10:711-717 (1998).

Atmanspacher, H.: Toward an information theoretical implementation of contextual conditions for consciousness. Acta Biotheoretica 54:157-160 (2006).

Avital, E., Jablonka, E.,: Animal Traditions: behavioral inheritance in evolution, Cambridge University Press, UK (2000).

Baars, B.: A Cognitive Theory of Consciousness. Cambridge University Press, New York (1988).

Baars, B.: Global workspace theory of consciousness: toward a cognitive neuroscience of human experience. Progress in Brain Research 150:45-53 (2005).

Backdahl, L., Bushell, A., Beck, S.: Inflammatory signalling as mediator of epigenetic modulation in tissue-specific chronic inflammation. The International Journal of Biochemistry and Cell Biology, doi:10.1016/j.biocel.2008.08.023 (2009).

Bailey, N.: The Mathematical Theory of Infectious Diseases and its Applications. Hafner: New York (1975).

Balkwill, F., Mantovani, A.: Inflammation and cancer: back to Virchow?, The Lancet, 357:539-545 (2001).

Bandelow, B., Spath, C., Tichauer,G., Broocks, A., Hajak, G., Ruther, E.: Early traumatic life events, parental attitudes, family history, and birth risk factors in patients with panic disorder, Comprehensive Psychiatry, 43:269-278 (2002).

Barker, D.: Fetal programming of coronary heart disease, Trends in Endocrinology and Metabolism, 13:364-372 (2002).

Barker D., Forsen, T., Uutela, A., Osmond, C., Erikson, J.: Size at birth and resilience to effects of poor living conditions in adult life: longitudinal study, British Medical Journal, 323:1261-1262 (2002).

Barkow J., Cosmides L., Tooby, J. eds.: The Adapted Mind: Biological Approaches to Mind and Culture, University of Toronto Press (1992).

Barnett, E., Halverson, J.: Disparities in premature coronary heart disease mortality by region and urbanicity among black and white adults ages 35-64, 1985-1995, Public Health Reports, 115:52-64 (2000).

Baverstock, K.: Radiation-induced genomic instability: a paradigm-breaking phenomenon and its relevance to environmentally induced cancer, Mutation Research, 454:89-109 (2000).

Beck C., Schlogl, F.: Thermodynamics of Chaotic Systems. Cambridge University Press, Cambridge, UK (1995).

Bennett, M., Hacker, P.: Philosophical Foundations of Neuroscience. Blackwell Publishing (2003).

Bennett, C.: Logical depth and physical complexity. In The Universal Turing Machine: A Half-Century Survey, R. Herkin (ed.), pp. 227-257, Oxford University Press (1988).

Ben-Tovin, D., Dougherty, M., Stapelton, A., Pinnock C.: Coping with prostate cancer: a Quantitative analysis using a new instrument, the center for clinical excellence in Urological research coping with cancer instrument, Urology 59:383-388 (2002).

Bertram, J.: The molecular biology of cancer, Molecular Aspects of Medicine, 21:167-223 (2001).

Bigsby, R., Chapin R., Daston G., Davis B., Gorski J., Gray L., Howdeshell K., Zoeller R., Vom Saal F.: Evaluating the effects of endocrine disruptors on Endocrine function during development. Environmental Health Perspectives, 107(supp 4):613-618 (1999).

Binney J., Dowrick, N., Fisher, A., Newman, M.: The theory of critical phenomena, Clarendon Press, Oxford, UK (1986).

Bjorntorp P.: Do stress reactions cause abdominal obesity and comorbidities?, Obesity Reviews, 2:73-86 (2001).

Blake, G., Ridker, P.: Inflammatory bio-markers and cardiovascular risk prediction, Journal of Internal Medicine, 252:283-294 (2001).

Blake, G., Ridker P.: Inflammatory mechanisms in atherosclerosis, Italian Heart Journal, 2:796-800 (2001).

Bongu, A., Chang, E., Ramsey-Goldman, R.: Can morbidity and mortality in SLE be improved? Best Practice and Research in Clinical Rheumatology 16:313-332 (2002).

Bonner J.: The evolution of culture in animals, Princeton University Press, Princeton, NJ (1980).

Bootsma, D., Hoeijmakers, J.:DNA repair: engagement with transcription, (news: comment), Nature, 363:114-115 (1993).

Bornstein, S., Licinio, J., Tauchnitz, R., Engelmann, L., Negrao, A., Gould, P., Chrousos, G.: Plasma leptin levels are increased in survivors of acute sepsis: associated loss of diurnal rhythm, in cortisol and leptin secretion, Journal of Clinical Endicrinolgy and Metabolism, 83:280-283 (1998).

Bos, R.: Continuous representations of groupoids. arXiv:math/0612639 (2007).

Bosma, H., Marmot, M., Hemingway, H., Nicholson, A., Brunner, E., Stansfeld, S: Low job control and risk of coronary heart disease in Whitehall II (prospective cohort) study, British Medical Journal, 314:558-565 (1997).

Bosma, H., Stansfeld, S., Marmot, M.: Job control, personal characteristics, and heart disease, Journal of Occupational Health Psychology, 3:402-409 (1998).

Bosma, H., Peter, R., Siegrist, J., Marmot, M.: Two alternative job stress models and the risk of coronary heart disease, American Journal of Public Health, 88:68-74 (2001).

Bossdorf, O., Richards, C., and Pigliucci, M.: Epigenetics for ecologists. Ecology Letters 11: 106-115 (2008).

Braiman Y., Linder J., Ditto, W.: Taming spatiotemporal chaos with disorder Nature, 378:465-469 (1995).

Britten, R., Davidson, E.: Gene regulation for higher cells: a theory. Science 165:349-357 (1969).

Brooks, A., Byrd, R., Weitzman, M., Auinger, P., McBride, J.: Impact of low birth weight on early childhood asthma in the United States. Archives of Pediatric and Adolescent Medicine, 155:401-406 (2001).

Brown, R.: From groups to groupoids: a brief survey. Bulletin of the London Mathematical Society 19:113-134 (1987).

Brunner, E., Marmot, M., Nanchahal, K., Shipley, M., Stansfeld, S., Juneja M., Alberti, K.: Social inequality in coronary risk: central obesity and the metabolic syndrome. Evidence from the Whitehall II study, Diabetologia, 40:1341-1349, (1997).

Brunner, E. et al., (16 authors): Adrenocortical, autonomic, and inflammatory causes of the Metabolic syndrome: nested case-control study. Circulation, 106:2659-2665 (2002).

Bullo, M., Garcia-Lorda, P., Megias, I., Salas-Salvado, J.: Systemic inflammation, adipose tissue tumor necrosis factor, and leptin expression. Obesity Research, 11:525.531 (2003).

Buneci, M.: Representare de Groupoizi. Editura Mirton, Timisoara (2003).

Bungum, T., Satterwhite, M., Jackson, A., Morrow, J. Jr.: The relationship of body mass index, medical costs, and job absenteeism. American Journal of Health Behavior, 27:456-462 (2003).

Bursik, R.: Delinquency rates as sources of ecological change. In:The Social Ecology of Crime (Byrne, J., Sampson, R., eds.) Springer-Verlag, New York (1986).

Cannas Da Silva, A., Weinstein A.: Geometric Models for Noncommutative Algebras. American Mathematical Society, RI (1999).

Carr W., Zeitel, L., Weiss, K.: Variations in asthma hospitalizations and deaths in New York City, American Journal of Public Health 82:59-65 (1992).

Casanueva, F., Dieguez, C.: Neuroendocrine regulation and actions of leptin. Frontiers in Neuroendocrinology, 20:317-363 (1999).

Caswell H.: Matrix Population Models, Aldine (2001).

Cayron, C.: Groupoid of orentational variants, Acta Crystallographica A, A62:21-40 (2006).

Cayron, C.,: Multiple twinning in cubic crystals: geometric/algebraic study and its application for the identification of the $\Sigma 3^n$ grain boundaries, Acta Crystallographica A, A63:11-29 (2007).

CDC, Centers for Disease Control: Asthma mortality and hospitalization among children and young adults – United States, 1980-93. Morbidity and Mortality Weekly Report, 45:350-353 (1996).

CDC: 1991-2001 Prevalence of Obesity Among US Adults, by Characteristics: Behavioral Risk Factor Surveillance System (1991-2001); Self-reported data, (2003).

http://www.cdc.gov/nccdphp/dnpa/obesity/trend/prev_char.htm

Chrousos, G.: The role of stress and hypothalamic-pituitary-adrenal axis in the pathogenesis of the metabolic syndrome: neuro-endocrine and target tissue-related causes, International Journal of Obesity and Related Metabolic Disorders, Suppl. 2:S50-S55, (2000).

Ciliberti, S., Martin, O., Wagner, A.: Robustness can evolve gradually in complex regulatory networks with varying topology. PLoS Computational Biology 3(2):e15 (2007).

Ciliberti, S., Martin, O., Wagner, A.: Innovation and robustness in complex regulatory gene networks. Proceedings of the National Academy of Sciences 104:13591-13596 (2007).

Cliff, A., Haggett, P., Ord, J.: Spatial Diffusion: An Historical Geography of Epidemics in an Island Community. Cambridge University Press, New York (1981).

Clougherty, J., Levy, J., Kubzansky, L., Ryan, P., Suglia, S., Canner, M., Wright, R.: Synergistic effects of traffic-related air pollution and exposure to violence on urban asthma etiology. Environmental Health Perspectives, 115:1140-1146 (2007).

Cohen I.: The cognitive principle challenges clonal selection, Immunology Today, 13:441-444 (1992).

Cohen I.: Tending Adam's Garden: Evolving the Cognitive Immune Self, Academic Press, New York (2000).

Cohen, I.: Immune system computation and the immunological homunculus. In Nierstrasz, O., Whittle, J., Harel, D., Reggio, G., (eds.), MoDELS 2006, LNCS, vol. 4199, pp. 499-512, Springer, Heidelberg (2006).

Cohen, I., Harel, D.: Explaining a complex living system: dynamics, multi-scaling, and emergence. Journal of the Royal Society: Interface 4:175-182 (2007).

Collins, C., Williams, D.: Examining the black-white adult mortality: the role of residential segregation, for 25th Public Health Conf. on Records and Statistics and the National Committee on Vital and Health Statistics 45th Anniversary Symp., July 17-19, 1996, Washington DC, (1996).

Collins, J. Jr., Schulte, N., Drolet, A.: Differnetial effect of ecological risk factors on the low Birthweight components of African-American, Mexican-American and non-Latino white infants in Chicago. Journal of the National Medical Association, 90:223-229 (1998a).

Collins, J. Jr., David, R., Symons, R., Handler, A., Wall, S., Andes, S.: African-American mother's perception of their residential environment, stressful life events, and very low birthweight. Epidemiology, 9:286-289 (1998b).

Cooke D.: Psychopathic personality in different cultures: what do we know? Journal of Personality Disorders, 10:23-40 (1996).

Cooper, R., Goldenberg, R., Das, A., Elder, N., Swain, M., Norman, G., Ramsey, R., Cotroneo, P., Collins, B., Johnson, F., Jones, P., Meier, A.: The preterm prediction study: maternal stress is associated with spontaneous preterm birth at less than thirty-five weeks' gestation. National Institute of Child Health and Human Development Maternal-Fetal Medicine Units Network. American Journal of Obstetrics and Gynecology, 175:1286-1292 (1996).

Cooper, R.: Trends and disparities in coronary heart disease stroke, and other cardiovascular diseases in the United States: findings of the National Conference on Cardiovascular Disease, Circulation, 102:3137-3147 (2000).

Cooper, R.: Social inequality, ethnicity and cardiovascular disease, Int J Epidem., 30(Supp 1):S48-52 (2001).

Coplan J., Altemus, M., Matthew, S., Smith, E., Scharf, B., Coplan, P., Kral, J., Gorman, J., Owens, M., Nemeroff, C., Rosenblum, L.: Synchronized maternal-infant elevations of primate CSF CRF concentrations in response to variable foraging demand, CNS Spectrums, 10:530-536 (2005).

Cosmides, L., Tooby, J.: Cognitive adaptations for social exchange. In The Adapted Mind: Evolutionary Psychology and the Generation of Culture, Oxford University Press, New York (1992).

Coussens, L., and Werb, Z.: Inflammation and Cancer, Nature, 420:860-867 (2002).

Cover, T., Thomas, J.: Elements of Information Theory. John Wiley and Sons, New York (1991).

Crews, D., and J. McLachlan, J.: Epigenetics, evolution, endocrine disruption, health, and disease. Endocrinology, 147:S4-S10 (2006).

Crews, D., Gore, A., Hsu, T., Dangleben, N., Spinetta, M., Schallert, T., Anway, M., Skinner, M.: Transgenerational epigenetic imprints on mate preference. Proceedings of the National Academy of Sciences 104:5942-5946 (2007).

Cross, Benton: The roles of interleukin-6 and interleukin-10 in B cell hyperactivity in systematic lupus erythematosus, Inflammation Research 48:255-261 (1999).

Crowther, N., Cameron, N., Trusler, J., Gray, I.: Association between poor glucose tolerance and rapid Postnatal weight gain in seven-year-old children. Diabetologia, 41:1163-1167 (1998).

Currie, E.: Reckoning, Hill and Wang (Farrar, Straus and Giroux), New York (1993).

Dalgleish, A.: The relevance of non-linear mathematics (chaos theory) to the treatment of cancer, the role of the immune response and the potential for vaccines, Quarterly Journal of Medicine, 92:347-359 (1999).

Dalgleish, A., O'Byrne, K.: Chronic immune activation and inflammation in the pathogenesis of AIDS and cancer, Advances in Cancer Research, 84:231-276 (2002).

Daly, M., Duncan, G., Kaplan, G., Lynch, J.: Macro-to-micro links in the relation Between income inequality and mortality. Milbank Qurterly, 76:315-339 (1998).

De Groot, J., Ruis, M., Scholten, J., Koolhass, J., Boersma, W.: Long-term effects of social stress on antiviral immunity in pigs, Physiological Behavior, 73:145-158 (2001).

Dehaene, S., Naccache, L.: Towards a cognitive neuroscience of consciousness: basic evidence and a workspace framework. Cognition 79:1-37 (2001).

Del Rio, C., Barragen, M.: Pneumocystis pneumonia remains important cause of Morbidity and mortality among inner city patients. New England Journal of Medicine, 351:1262-1263 (2004).

Dembo, A., Zeitouni, O.: Large Deviations: Techniques and Applications, 2nd edition. Springer, New York (1998).

DHPD (Department of Housing, Preservation, and Development): Summary of Housing, New York City 1996. New York City: DHPD. Available from The City Book Store and at the New York City government web site (1999).

Dias, A., Stewart, I.: Symmetry groupoids and admissible vector fields for coupled cell networks, Journal of the London Mathematical Society 69:707-736 (2004).

Dimitrov A., Miller, J.: Neural coding and decoding: communication channels and quantization, Network: Computation in Neural Systems, 12:441-472 (2001).

Dretske, F.: The explanatory role of information. Philosophical Transactions of the Royal Society A 349:59-70 (1994).

Drucker, E., Alcabes, P., Bosworth, W., Sckell, B.: Childhood tuberculosis in the Bronx, New York. The Lancet, 343:1482-1485 (1995).

Du Clos, T.: Function of C-reactive protein, Annals of Medicine, 32:274-278 (2000).

Durham W.: Coevolution: Genes, Culture, and Human Diversity, Stanford University Press, Palo Alto, CA (1991).

Duryea, P.: Press release dated Friday 27 January, 1978, Office of the New York State Assembly Republican Leader, Albany, NY. Copy available from R. Wallace, (1978).

Egger, G., Swinburn, B., An ecological approach to the obesity pandemic. British Medical Journal 315:477-480 (1997).

Egle U., Hardt, J., Nickel, R., Kappis, B., Hoffmann, S.: Long-term effects of adverse childhood experiences – Actual evidence and needs for research, Zeitschrift fur Psychosomatische Medizin und Psychotherapie, 48:411-434 (2002).

Eliman, A., Knutsson, U., Gronnegard, M., Steirna, A., Albertsson-Wiklan, K., Marcus, C.: Variations in glucocorticoid levels within the physiological range affect plasma leptin levels, European Journal of Endocrinology, 139:615-620 (1998).

Eldredge N.: Time Frames: The Rethinking of Darwinian Evolution and the Theory of Punctuated Equilibria, Simon and Schuster, New York (1985).

Emery, M.: Stochastic Calculus on Manifolds. Springer, New York (1989).

English, T.: Evaluation of evolutionary and genetic optimizers: no free lunch. In Fogel, L, P. Angeline, and T. Back (eds.), Evolutionary Programming V: Proceedings of the Fifth Annual Conference on Evolutionary Programming, 163-169, MIT Press, Cambridge, MA (1996).

Erdos, P., Renyi, A.: On the evolution of random graphs (1960). Reprinted in The Art of Counting, 1973, pp. 574-618 (1973), and in Selected Papers of Alfred Renyi, pp. 482-525 (1976).

Eriksson, J., Forsen, T., Tuomilehto, J., Winter, P., Osmond, C., Barker, D.: Catch-up growth in childhood and death from coronary heart disease: longitudinal study. British Medical Journal, 318:427-431 (1999).

Eriksson J., Forsen, T., Tuomilehto, J., Osmond, C., Barker, D.: Fetal and childhood growth and hypertension in adult life, Hypertension, 36:790-794 (2000).

Evan, G., Littlewood, T.: A matter of life and cell death, Science, 281:1317-1322 (1998).

Evans, R., Barer, M., Marmor, T. (eds): Why Are Some People Healthy and Others Not?: The Determinants of Health of Populations, Aldine De Gruyter: New York (1994).

Eyer, J.: Does unemployment cause the death rate peak in each business cycle? A multifactor model of death rate change, International Journal of Health Services, 7:625-662 (1977).

Fagan, J., Wilkinson, D.: Guns, youth violence, and social identity. In: Youth Violence (Crime and Justice, 24), eds. Tonry, M., Moore, R., pp. 235-250 (1999).

Fang, J., Madhavan, S., Bosworth, W., Alderman, M.: Residential segregation and mortality in New York City, Social Science and Medicine, 47:469-476 (1998).

Fichtner, K.: Non-space-group symmetry in crystallography, Computers and Mathematics with Applications, B, 12:751-762 (1986).

Ficici, S., Milnik, O., Pollak, J.: A game-theoretic and dynamical systems analysis of selection methods in coevolution. IEEE Transactions on Evolutionary Computation 9:580-602 (2005).

Fitz, D., Reiner, H., Plakensteiner, K., Rode, B.: 2007, Possible origins of biohomochirality, Current Chemical Biology, 1:41-52 (2007).

Feynman, R.: Lectures on Computation, Westview Press, New York (2000).

Flegal, K., Carroll, M., Ogden, C., Johnson, C.: Prevalence and trends in obesity among US adults 1999-2000. Journal of the American Medical Association, 288:1723-1727 (2002).

Fleming R., Shoemaker, C.: Evaluating models for spruce budworm-forest management: comparing output with regional field data, Ecological Applications, 2:460-477 (1992).

Fodor J.: The Language of Thought, Corwell, New York, (1975).

Foley, D., Craid, J., Morley, R., Olsson, C., Dwyer, T., Smith, K., Saffery, R.: Prospects for epigenetic epidemiology. American Journal of Epidemiology 169:389-400 (2009).

Fontainer, K., Redden, D., Wang, C., Westfall, A., Allison, D.: 2003. Years of life lost to obesity. Journal of the American Medical Association, 289:187-193 (2003).

Forlenza M., Baum, A.: Psychosocial influences on cancer progression: alternative cellular and molecular mechanisms, Current Opinion in Psychiatry, 13:639-645 (2000).

Fried, S., Ricci, M., Russell, C., Laferrere, B.: Regulation of leptin production in humans, Journal of Nutrition, 130:3127-3131 (2000).

Fullilove, M.: Root Shock: Upheaval, Resettlement and Recovery in Urban America, Random House, New York, (2004).

Fullilove, M., Fullilove, R.: Intersecting epidemics: black teen crack use and sexually transmitted disease. Journal of the American Medical Association, 44:146-153 (1989).

Galea, S., Ahern, J., Vlahov, D., Coffin, P., Fuller, C., Leon, A., Tardiff, K.: Income distibution and risk of fatal drug overdose in New York City neighborhoods, Drug and Alcohol Dependency, 70:139-148 (2003).

Gallagher, R., Marbach, J., Raphael, K., Handte, J., Dohrenwend, B.: Myofascial face pain: seasonal variability in pain intensity and demoralization, Pain, 61:113-120 (1995).

Gammaitoni A., Hanggi P., Jung, P. and F. Marchesoni: Stochastic resonance, Reviews of Modern Physics, 70:223-287 (1998).

Geronimus A., Bound J., Waidmann T.: Poverty, time, and place: Variation in excess mortality across selected US populations, 1980-1990, Journal of Epidemiology and Community Health, 53:325-334 (1999).

Gilbert, P.: Evolutionary approaches to psychopathology: the role of natural defenses, Australian and New Zealand Journal of Psychiatry, 35:17-27 (2001).

Gilbert, S.: Mechanisms for the environmental regulation of gene expression: ecological aspects of animal development. Journal of Bioscience 30:65-74 (2001).

Ginsberg-Fellner F., Jagendorf, L., Carmel, H., Harris, T.: Overweight and obesity in preschool children in New York City, American Journal of Nutrition, 34:2236-2241 (1981).

Giovambattista, A., Chisari, A., Gaillard, R., Spinedi, E.: Food intake-induced leptin secretion modulates hypothalamo-pituitary-adrenal axis response and hypothalamic Ob-Rb expression to insulin administration. Neuroendocrinology, 72:341-349 (2000).

Glazebrook, J.F., R. Wallace, R.: Small worlds and red queens in the global workspace: an information-theoretic approach. Cognitive Systems Research, 10:333-365 (2009).

Gleiser, M., Thorarinson, J., Walker, S.: Punctuated chirality, Origins of Life and Evolution of Biospheres, DOI 10.1007/s11084-008-9147-0 (2008).

Godfrey, K., Barker, D.: Fetal programming and adult health, Public Health and Nutrition, 4:611-624 (2001).

Golubitsky, M., Stewart, I.: Nonlinear dynamics and networks: the groupoid formalism. Bulletin of the American Mathematical Society 43:305-364 (2006).

Gong, G., Oakley-Girvan, I., Wu, A., Kolonel, L., John, E., West, D., Felberg, A., Ballagher, R., Whittemore A.: Segregaton analysis of prostate cancer in 1719 white, African-American and Asian-American families in the US and Canada, Cancer Causes and Control, 13:471-482 (2002).

Goran, M., Ball, G., Cruz, M.: Obesity and risk of type 2 diabetes and cardiovascular diseases in Children and adolescents. Journal of Clinical Endocrinology and Metabolism, 88:1417-1427 (2003).

Goubault, E., Raussen M.: Dihomotopy as a tool in state space analysis. Lecture Notes in Computer Science, Vol. 2286, April, 2002, 16-37, Springer, New York (2002).

Goubault, E.: Some geometric perspectives on concurrency theory. Homology, Homotopy, and Applications 5:95-136 (2003).

Gould, P.: The Slow Plague. Blackwell, London (1993).

Gould, S.,: The Structure of Evolutionary Theory. Harvard University Press, Cambridge, MA (2002).

Granovetter, M.: The strength of weak ties, American Journal of Sociology, 78:1360-1380 (1973).

Greenlund, K., Giles, W., Keenan, N., Croft, J., Casper, M., Matson-Koffman, D.: 1998. Prevalence of multiple cardiovascular disease risk factors among women in the United States,1992 and 1995: the Behavioral Risk Factor Surveillance System. Women's Health, 7:1125-1133 (1998).

Griscom, J.: The Sanitary Condition of the Laboring Population of New York, Arno Press, New York (1844, reprinted 1970).

Grossi, G., Ahs, A., Lundberg, U.: Psychological correlates of salivary cortisol secretion among unemployed men and women, Integrated Physiology and Behavioral Science, 33:249-263 (1998).

Grossman, Z.: The concept of idiotypic network: deficient or premature? In: H. Atlan and IR Cohen, (eds.), Theories of Immune Networks, Springer Verlag, Berlin, p. 3852 (1989).

Grossman, Z.: Contextual discrimination of antigens by the immune system: towards a unifying hypothesis, in: A. Perelson and G. Weisbch, (eds.) Theoretical and Experimental Insights into Immunology, Springer Verlag, p. 7189 (1992a).

Grossman, Z.: International Journal of Neuroscience, 64:275 (1992b).

Grossman, Z.: Cellular tolerance as a dynamic state of the adaptable lymphocyte, Immunology Reviews, 133:45-73 (1993).

Grossman, Z.: Round 3, Seminars in Immunology, 12:313-318 (2000).

Gryazeva, A., Shurlygina, I., Verbitskaya, L., Mel'nikova, E., Kudryavtseva, N., Trufakin, V.: Changes in various measures of immune status in mice subject to social conflict, Neuroscience and Behavioral Physiology, 31:75-81 (2001).

Guerrero-Bosagna, C., Sabat, P., Valladares, L.: Environmental signaling and evolutionary change: can exposure of pregnant mammals to environmental estrogens lead to epigenetically induced evolutionary changes in embryos?. Evolution and Development 7:341-350 (2005).

Guinan, M.: Black communities' belief in 'AIDS as genocide'. A barrier to overcome for HIV prevention. Annals of Epidemiology, 3:193-195 (1993).

Gunderson, L.: Ecological resilience-in theory and application, Annual Reviews of Ecological Systematics, 31:425-439 (2000).

Harris, M.: Diabetes in America: epidemiology and scope of the problem. Diabetes Care, 21(suppl. 3):C11-C14 (1998).

Hartl, D., Clark, A.: Principles of Population Genetics, Sinaur Associates, Sunderland, MA, (1997).

Heneghan, C., Chow, C., Collins, J., Imhoff, T., Lowen, S., Teich, M.: Information measures quantifying aperiodic stochastic resonance, Physical Review A, 54:2366-2377 (1996).

Herberman R.: Principles of tumor immunology in Murphy, G., Lawrence, W., Lenhard, R. (eds.), American Cancer Society Textbook of Clinical Oncology, ACS, Second Edition, pp. 1-9, (1995).

Herpertz, S., Sass H.: Emotional deficiency and psychopathy, Behavioral Sciences and the Law, 18:567-580 (2000).

Hershberg, U., Efroni, S.: The immune system and other cognitive systems, Complexity, 6:14-21 (2001).

HHS (Health and Human Services): Health, United States, 1998, with Socioeconomic Status and Health Chartbook, Washington, D.C. (1998).

Hill, J., Wyatt, H., Reed, G., Peters, J.: Obesity and the environment: where do we go from here? Science, 266:853-858, 2003.

Hillemeier, M., Lynch, J., Harper, S., Raghunathan, T., Kaplan, G.: Relative or absolute standards for child poverty: a state-level analysis of infant and child mortality, American Journal of Public Health, 93:652-657 (2003).

Hilty C., Bruhlmann, P., Sprott, H., Gay, R., Michel, B., Gay, S., Neidhart, M.: Altered diurnal rhythm of prolactin in systemic sclerosis, Rheumatology 27:2160-2165 (2000).

Hirsch, J.: Obesity: matter over mind? Cerebrum, 5:7-18 (2003).

Hoffman-Goetz, L., Mills, S.: Cultural barriars to cancer screening among African American women: A critical review of the qualitative literature. Women's Health, 3:183-201 (1997).

Holling, C.: Resilience and stability of ecological systems, Annual Reviews of Ecological Systematics, 4:1-23 (1973).

Holling, C.: Cross-scale morphology, geometry and dynamicsl of ecosystems. Ecological Monographs 41:1-50 (1992).

Houseknecht, K., Baile, C., Matteri, R., Spurlock, M.: The biology of leptin: a review. Journal of Animal Science, 76:1405-1420 (1998).

Hunt, L., Chambers, C.: The Heroin Epidemics, Spectrum Publications (John Wiley and Sons), New York, (1976).

Iqbal, J., Pompolo, S., Murakami, R., Clarke, I.: Localization of long-form leptin in the somatostatin-containing neurons in the sheep hypothalamus, Brain Res., 887:1-6 (2000).

Ives, A.: Measuring resilience in stochastic systems, Ecological Monographs, 65:217-233, (1995).

Jablonka, E., and Lamb, M.: Epigenetic Inheritance and Evolution: The Lamarckian Dimension. Oxford University Press, Oxford, UK (1995).

Jablonka, E., M. Lamb, M.: Epigenetic inheritance in evolution. Journal of Evolutionary Biology 11:159-183 (1998).

Jablonka, E.: Epigenetic epidemiology. International Journal of Epidemiology 33:929-935 (2004).

Jaeger, J., Surkova, S., Blagov, M., Janssens, H., Kosman, D., Kozlov, K., Manu, M., Myasnikova, E., Vanario-Alonso, C., Samsonova, M., Sharp, D., Reintiz, J.: Dynamic control of positional information in the early Drosophila embryo. Nature 430:368-371 (2004).

Jaenisch, R., A. Bird, A.: Epigenetic regulation of gene expression: how the genome integrates intrinsic and environmental signals. Nature Genetics Supplement 33:245-254 (2003).

Jones, I., Blackshaw, J.: An evolutionary approach to psychiatry, Australian and New Zealand Journal of Psychiatry, 34:8-13 (2000).

Kadtke, J., Bulsara, A.: Applied Nonlinear Dynamics and Stochastic Systems Near the Millenium, AIP Conference Proceedings, American Institute of Physics, New York (1997).

Kane, G.: Inner city alcoholism: an ecological analysis and cross-cultural study. Human Services Press, New York, (1981).

Kaplan, J., Adams, M., Clarkson, T., Manuck, S., Shively, C., Williams, J.: Psychosocial factors, sex differences, and atherosclerosis: lessons from animal models Psychosocial Medicine, 58:598-611 (1996).

Karlsen, S., Nazroo, J.: Relation between racial discrimination, social class, and health among ethnic minority groups, American Journal of Public Health, 92:624-631 (2002).

Kastner, M.: Phase transitions and configuration space topology. ArXiv cond-mat/0703401 (2006).

Kawachi, I., Kennedy, B., Lochner, K., Prothrow-Stith, D.: Social capital, income inequality, and mortality, Amer. J. Public Health, 87:1491-1498 (1997).

Khinchin, A.: Mathematical Foundations of Information Theory. Dover, New York (1957).

Kiecolt-Glaser, J., McGuier, L., Robles, T., Glaser, R.: Emotions, morbidity, and mortality: new perspectives from psychoneuroimmunology, Annual Review of Psychology, 53:83-107 (2002).

Kimm, S., Obarzanek, E., Childhood obesity: a new pandemic of the new millennium. Pediatrics: 110:1003-1007 (2002).

Kivimaki, M., Leino-Arjas, P., Luukkonen, R., Riihimaki, H., Vahtera, J., Kirjonen, J.: Work stress and risk of cardiovascular mortality: prospective cohort study of industrial employees, British Medical Journal, 325:456-466 (2002).

Klemperer, W.: The steric courses of chemical reaction, Journal of the American Chemical Society, 95:380-396 (1973).

Krebs, P.: Models of cognition: neurological possibility does not indicate neurological plausibility. In Bara, B., L. Barsalou, Bucciarelli, M. (eds.). Proceedings of CogSci 2005, 1184-1189, Stresa, Italy.
Available at http//cogprints.org/4498/ (2005).

Lalumiere, M., Harris, G., Rice, M.: Psychopathy and developmental instability, Evolution and Human Behavior, 22:75-92 (2001).

Landau, L., Lifshitz, E.: Statistical Physics, 3rd Edition, Part I. Elsevier, New York (2007).

Latkin, C., Knowlton, A., Hoover D., Mandell, W.: Drug network characteristics as a predictor of cessation of drug use among adult injection drug users: a prospective study, American Journal of Drug and Alcohol Abuse, 25:463-473 (1999).

Lechner O., Dietrich, H., Oliveria dos Santos, A., Wiegers, G., Schwarz S., Harbutz, M., Herold, M., Wick, G.: Altered circadian rhythms of the stress hormone and melatonin response in lupus-prone MLR/MP-fas(lpr) mice, Journal of Autoimmunity 14:325-333 (2000).

Lee, J.: Introduction to topological manifolds. Springer, New York (2000).

Lewontin, R.: The Triple Helix: Gene, Organism and Environment. Harvard University Press, Cambridge, MA. 2000.

Li, X., Howard, D., Stanton, B., Rachuba, L., Cross, S.: Distress symptoms among urban African American children and adolescents: a psychometric evaluation of the Checklist of Children's Distress Symptoms. Archives of Pediatric and Adolescent Medicine, 152:569-577 (1998).

Li, X., Standton, B., Feigelman, S.: Exposure to drug trafficking among urban, low-income African American children and adolescents. Archives of Pediatric and Adolescent Medicine, 153:161-168 (1999).

Liang M., Mandl, L., Costenbader, K., Fox, E., Karlson, E.: Atherosclerotic vascular disease in systemic lupus erythematosus, Journal of the National Medical Association 94:813-819 (2002).

Libbey P., Ridker, P., Maseri, A.: Inflammation and atherosclerosis, Circulation, 105:1135-1143 (2002).

Lindsay, R., Cook, V., Hanson, R., Salbe, A., Tataranni, A., Knowler, W.: Early excess weight gain of children in the Pima Indian population. Pediatrics, 109:E33 (2002).

Link, B., Phelan, J.: Evaluating the fundamental cause explanation for social disparities in health, Bird, C., Conrad, P., and Fremont, A., eds. Handbook of Medical Sociology, Fifth ed. Prentice-Hall, New Jersey, (2000).

Linker-Israeli M., Deans, R., Wallace, D., Prehn, J., Ozeri-Chen, T., Klinenberg, J.: Journal of Immunology, 147:117-123 (1991).

Liteonjua, A., Carey V., Weiss S., Gold, D.: Race, Socioeconomic factors, and area of residence are associated with asthma prevalence, Pediatric Pulmonology, 28:394-401 (1999).

Lord, G., Matarese, G., Howard, J., Baker, R., Bloom, S., Lechler, R.: Leptin modulates the T-cell immune response and reverses starvation-induced immunosuppression, Nature, 394:897-901 (1998).

Lottenberg, S., Giannella-Neto, D., Derendorf, H., Rocha, M., Bosco, M., Bosco, A., Carvalho, S., Moretti, A., Lerario, A., Wajchenberg, B.: Effect of fat distribution on the pharmacokinetics of cortisol in obesity. International Journal of Clinical Pharmacology Therapeutics, 36:501-505 (1998).

Maas, W., Natschlager, T., Markram, H: Real-time computing without stable states: a new framework for neural computation based on perturbations. Neural Computation 14, 2531-2560 (2002).

Magnasco, M. Thaler, D.: Changing the pace of evolution, Physics Letters A, 221:287-292 (1996).

Marmot, M., Bosma, H., Hemingway, H., Brunner, E., Stansfeld, S.: Contribution of job control and other risk factors to social variations in coronary heart disease incidence, The Lancet, 350:235-239 (1997).

Marmot, M., Fuhrer, G., Ettner, S., Marks, N., Bumpass, L., Ryff, C.: Contribution of psychosocial factors to socioeconomic differences in health. Milbank Quarterly, 76:403-448, (1998).

Massey, D.: American apartheid: segregation and the making of the underclass, American Journal of Sociology 96:329-357 (1990).

Massey, D., Denton, N.: American Apartheid, Harvard University Press (1992).

Matsumoto, Y.: An Introduction to Morse Theory. American Mathematical Society, Providence, RI (2002).

Maturana, H., F. Varela, F.: Autopoiesis and Cognition. Reidel Publishing Company, Dordrecht, Holland (1980).

Maturana, H., F. Varela, F.: The Tree of Knowledge. Shambhala Publications, Boston, MA (1992).

McCauly, J.: Chaos, Dynamics, and Fractals. Cambridge Nonlinear Science Series, Cambridge, UK (1994).

McClintock, N., Luchinsky, D.: Glorious noise, The New Scientist, 161(2168):36-39 (1999).

McCord C., Freeman H.: Excess mortality in Harlem, New England Journal of Medicine, 322:173-177 (1990).

McFarland, W., Chen, S., Hsu, L., Schwarcz, S., Katz, M.: 2003. Low socioeconomic status is Associated with a higher rate of death in the era of highly active antiretroviral Therapy, San Francisco, Journal of Acquirred Immune Deficinecy Syndrome, 33:96-103 (2003).

Mealey, L.: The sociobiology of sociopathy: an integrated evolutionary model, Behavioral and Brain Sciences, 18:523-599 (1995).

Melman, S.: The War Economy of the United States, St. Martin's Press, New York (1971).

Melnik, T., Rhoades, S., Wales, K., Cowell, C., Wofle, W. 1998. Overweight school children in New York City:Prevalence estimates and characteristics. Journal of Obesity and Related Metabolic Disorders, 22:7-13 (1998).

Miller, G., Chen, E.: Life stress and diminished expression of genes encoding glucocorticoid receptor and β_2-adrenergic receptor in children with asthma. Proceedings of the National Academy of Sciences, 103:5496-5501 (2006).

Mjolsness, E., Sharp, D., Reinitz, J.: A connectionist model of development. Journal of Theoretical Biology 152:429-458 (1991).

MMWR: Trends in death from systemic lupus erythematosus–United States, 1979-1998, Morbidity and Mortality Weekly Reports 51:371-374 (2002).

Mokdad, A., Serdula, M., Dietz, W., Bowman, B., Marks, J., Koplan, J.: The spread of the obesity epidemic in the United States, 1991-1998, Journal of the American Medical Association, 282:1519-1522, (1999).

Mokdad, A., Serdula, M., Dietz, W., Bowman, B., Marks, J., Koplan, J.: The continuing epidemic of obesity in the United States. Journal of the American Medical Association, 284:1650-1651 (2000).

Nakamura, K., Shimai, S., Kikuchi, S., Takahashi, H., Tanaka, M., Nakano, S., Motohashi, Y., Nakadaira, H., Yamamoto, M.: Increases in body mass index and waist circumference as outcomes of working overtime. Occupational Medicine (Lond), 48:169-173 (1998).

NCCP: Low income children in the United States, http://www.nccp.org/pub_cpf03.html (2003).

NCHS, National Center for Health Statistics: Medical care for asthma increasing and changing, Public Health Reports 87:112 (1996).

Neumark-Sztainer, D., Story, M., Falkner, N., Beuhring, T., Resnick, M.: Sociodemographic and personal characteristics of adolescents engaged in weight loss and weight/muscle gain behaviors: who is Doing what? Preventive Medicine, 28:40-50 (1999).

Newcomer, J., Selke, G., Melson, H., Gross, J., Vogler, G., Dagogo-Jack, S.: Dose-dependent cortisol-induced increases in plasma leptin concentration in healthy humans. Archives of General Psychiatry, 55:995-1000 (1998).

Nies, M., Vollman, M., Cook, T.: African Ameriocan women's experiences with physical activity in their daily lives. Public Health Nursing, 16:23-21 (1999).

Nisbett, R., Peng, K., Incheol, C., Norenzayan, A.: Culture and systems of thought: holistic vs. analytic cognition, Psychological Review, 108:291-310 (2001).

Nourse, J.: An algebraic description of stereochemical correspondence, Proceedings of the National Academy of Sciences, 72:2385-2388 (1975).

Nunney, L.: Lineage selection and the evolution of multistage carcinogenesis, Proceedings of the Royal Society, B, 266:493-498 (1999).

Nyirenda M., Seckl J.: Intrauterine event and the programming of adulthood disease: the role of fetal glucocorticoid exposure, International Journal of Molecular Medicine, 2:607-614 (1998).

O'Byrne, K., Dalgleish A.: Chronic immune activation and inflammation as the cause of malignancy, British Journal of Cancer, 85:473-483 (2001).

O'Campo P., Xue X., Wang M., O'Brien Caughey M.: Neighborhood risk factors for low birthweight in Baltimore: a multilevel analysis, American Journal of Public Health, 87:1113-1118 (1997).

Office of the Public Advocate: Food for Thought: How the Food Industry Can Help End Hunger in NYC.
http://publicadvocate.nyc.gov.policy/food$_{f}or_{t}hought.html$(2003).

Ofria, C., Adami, C., Collier, T.: Selective pressures on genomes in molecular evolution, Journal of Theoretical Biology, 222:477-483 (2003).

Ogden, C., Troiano, R., Briefel, R., Kuczmarski, R., Flegal, K., Johnson, C.: Prevalence of overweight among preschool children in the United States, 1971-1994. Pediatrics, 99:E1 (1997).

O'Nuallain, S.: Code and context in gene expression, cognition, and consciousness. Chapter 15 in Barbiere, M. (ed.), The Codes of Life: The Rules of Macroevolution. Springer, New York, pp. 347-356 (2008).

O'Nuallain, S., and R. Strohman, R.: Genome and natural language: how far can the analogy be extended? In Witzany (ed.), Proceedings of Biosemiotics. Tartu University Press, Umweb, Finland (2007).

Osmond, C., Barker D.: Environmental Health Perspectives, 108, Suppl. 3:545-553 (2000).

Ostlin, P., Diderichsen, F.: Policy Learning Curve Series. Equity-Oriented National Strategy for Public Health in Sweden. A Case Study. WHO Europe: Brussels (2001).

Pappas, G.: The Magic City, Cornell University Press, Ithaca, NY, (1989).

Paris, J.: Personality disorders: a biopsychosocial model, Journal of Personality Disorders, 7:53-99 (1993).

Parker, S., Davis K., Wingo P., Ries L., Health C.: Cancer statistics by race and ethnicity, Canadian Cancer Journal 48:31-49 (1998).

Parks C., Cooper, G.: Explaining racial disparity in systemic lupus erythematosus: environmental and genetic risk factors in the Carolina lupus study, Abstracts of the AEP 12:502 (2002).

Pattee, H.: Laws and constraints, symbols and languages, in C. Waddington (ed.) Towards a Theoretical Biology: Essays, Aldine-Atherton, Chicago (1972).

Paul W.: Fundamental Immunology, 4th Edition, Lippincott, Williams and Wilkins, New York (1999).

Perrotta, A., Jones, C., Braconi, F., Toribio, E., Stern P.: Out of Balance: The Housing Crisis from a Regional Perspective. The Regional Plan Association. http://www.rpa.org, (2004)

Pettini, M.: Geometry and Topology in Hamiltonian Dynamics and Statistical Mechanics. Springer, New York (2007).

Phillips, D.: Birthweight and the future development of diabetes: A review of the evidence, Diabetes Care, 21(supp. 2):150-155 (1998).

Phillips, D., Fall, C., Cooper, C., Norman, R., Robinson, J., Owens, P.: 1999. Size at birth and plasma leptin concentrations in adult life. International Journal of Obesity and Related Metabolic Disorders, 23:1025-1029 (1999).

Pielou, E.: Population and Community Ecology: Principles and Methods, Gordon and Beach, New York (1976).

Pielou, E.: Mathematical Ecology, John Wiley and Sons, New York, (1977).

Podolsky, S., Tauber, A.: The generation of diversity: Clonal selection theory and the rise of molecular biology, Harvard University Press (1997).

Polednak, A.: Black-white differences in infant mortality in 38 SMSA, American Journal of Public Health, 81:1480-1482 (1991).

Polednak, A.: Poverty, residential segregation, and black/white mortality ratios in urban areas, Journal of Health Care for the Poor and Underserviced, 4:363-373 (1993).

Polednak, A.: Segregation, discrimination and mortality in US Blacks, Ethniciey and Disease, 6:99-108 (1996).

Pratt, V.: Modeling concurrency with geometry. Proceedings of the 18th ACM SIGPLAN-SIGACT Symposium on Principles of Programming Languages, 311-322 (1991).

Rajaram S., Vinson V.: African American women and diabetes: a sociocultural context, Health Care of the Poor and Underserved, 9:236-247 (1998).

Rau, H., Elbert, T.: Psychophysiology of arterial baroreceptors and the etiology of hypertension, Biological Psychology, 57:179-201 (2001).

Reinitz, J., Sharp, D.: Mechanisms of even stripe formation. Mechanics of Development 49, 133-158 (1995).

Repetti, R.,Taylor, S., Seeman, T.: Risky families: family social environments and the mental and physical health of offspring, Psychological Bulletin, 128:330-366 (2002).

Richerson, P., Boyd R.: The evolution of human hypersociality, Paper for Rindberg Castle Symposium on Ideology, Warfare, and Indoctrination, (January, 1995), and HBES meeting, (1995).

Richerson, P., Boyd, R.: Not by Genes Alone: How Culture Transformed Human Evolution, Chicago University Press (2004).

Ridker, P.: On evolutionary biology, inflammation, infection, and the causes of atherosclerosis, Circulation, 105:2-4 (2002).

Ridley, M.: Evolution, Second Edition, Blackwell Science, Oxford, UK, (1996).

Roberts, E.: Neighborhood social environmental and the distribution of low birthweight in Chicago. American Journal of Public Health, 87:597-603 (1997).

Rodriguez-Sanz, M., Borrell, C., Urbanos, R., Pasarin, M., Rico, A., Fraile, M., Ramos, X., Navarro, V.: Power relations and premature mortality in Spains autonomous communities. International Journal of Health Servces, 33:687-722 (2003).

Rojdestvensky, I., Cottam, M.: Mapping of statistical physics to information theory with applications to biological systems, Journal of Theoretical Biology, 202:43-54 (2000).

Rose, N.: Mechanisms of autoimmunity, Seminars in Liver Disease 22:387-394 (2002).

Rosmond, R., Bjorntorp, P.: Endocrine and metabolic aberrations in men with abdominal obesity in relation to anxio-depressive infirmity. Metabolism 47:1187-1193 (1998).

Rosmond, R., Bjorntorp, P.: Psychosocial and socio-economic factors in women and their relationship to obesity and regional body fat distribution. International Journal of Obesity and Related Metabolic Disorders, 23:138-145 (1999).

Rosner, D. (ed.): Hives of Sickness. Rutgers University Press, New Brunswick (1995).

Ross, L.: Sterilization and 'de facto' sterilization. Amicus Journal, Winter: 29 (1994).

Ross, N., Wolfson, M., Dunn, J., Berthelot, J., Kaplan, G., Lynch, J.: Relation between Income inequality and mortality in Canada and in the United States: cross sectional Assessment using census data and vital statistics. British Medical Journal, 320:898-902 (2000).

Roth, J., Qiang, X., Marban, S., Redelt, H., Lowell, B.: The obesity pandemic: where have we been and where are we going. Obesity Research 12:88S-101S (2004).

Sadanaga, R., Ohsumi K.:, Basic theorems of vector symmetry in crystallography, Acta Crystallographica A, A35:115-122 (1979).

Sampson, R.: The impact of housing policies on community social disorganization and crime. Bulletin of the New York Academy of Medicine, 66:526-533 (1990).

Sampson R., Raudenbush S., Earls F.: Neighborhood and violent crime: a multilevel study of collective efficacy, Science, 277:918-924 (1997).

Sandars, P.: A toy model for the generation of homochirality during polymerization, Origins of Life and Evolution of Bioshperes, 33:575-587 (2003).

Santiago, E.: When I Was Puerto Rican. Vintage Books, New York, pp. 63-83 (1994).

Santos-Alvarez, J., Governa, R., Sanchez-Margalet, V.: Human leptin stimulates proliferation and activation of circulating monocytes, Cellular Immunology, 194:6-11 (1999).

Sarma, A., Schottenfeld, D.: Prostate cancer incidence, mortality, and survival trends in the United States,: 1981-2001, Seminars in Urological Oncology, 20:3-9 (2002).

Schachter, L., Salome, C., Peat, J., Woolcock, A.: Obesity is a risk for asthma and wheeze but not airway hyperresponsiveness. Thorasics, 56:4-8 (2001).

Schapiro, S., Nebete, P., Perlman, J., Bloomsmith, M., Sastry, K.: Effects of dominance status and environmental enrichment on cell-mediated immunity in rhesus macaques, Applied Animal Behavorial Science, 56:319-332 (1998).

Scherrer, K., J. Jost, J.: The gene and the genon concept: a functional and information-theoretic analysis. Molecular Systems Biology 3:87-95 (2007).

Scherrer, K., Jost, J.: Gene and genon concept: coding versus regulation. Theory in Bioscience 126:65-113 (2007).

Schotte, H., Schluter, B., Rust, S., Assmann, G., Domschke, W., Gaubitz, M.: Interleukin-6 promoter polymorphism (-174 G/C) in Caucasian German patients with systematic lupus erythematosus, Rheumatology 40:393-400 (2001).

Schubert C., Lampe A., Rumpold G., Fuchs D., Konig P., Chamson E, and Schussler, G.: Daily psychosocial stressors interfere with the dynamics of urine neopterin in a patient with systemic lupus erythematosus: an integrative single-case study, Psychosomatic Medicine 61:876-882 (1999).

Schultz, E.: Resolving the anti-antievolutionism dilemma: a brief for relational evolutionary thinking in anthropology. American Anthropologist, 111:224-237 (2009).

Schwartz, E., Granger, D., Susman, E., Gunnar, M., Laird, B.: Assessing salivary cortisol in studies of child development. Child Development, 69:1503-1513 (1998).

Shankardass, K., McConnell, R., Jerrett, M., Milam, J., Richardson, J., Berhane, K.: Parental stress increases the effect of traffic-related air pollution on childhood asthma incidence. Proceedings of the National Academy of Sciences, 106:12406-12411 (2009).

Sharp, D., Reinitz, J.: Prediction of mutant expression patterns using gene circuits. BioSystems 47:79-90 (1998).

Shavers, V., Harlan, L., Stevens, J.: Racial/ethnic variation in clinical presentation, treatment, and survival among breast cancer patients under age 35, Cancer, 97:134-147 (2003).

Shirkov, D., Kovalev, V.: The Bogoliubov renormalization group and solution symmetry in mathematical physics, Physics Reports, 352:219-249 (2001).

Shumow L., Vandell D., Posner J.: Perceptions of danger: a psychological Mediator of neighborhood demographic characteristics, American Journal of Orthopsychiatry, 68:468-478 (1998).

Singhal, A., Farooque, I., Cole, T., O'Rahilly, S., Fewtress, M., Kattenhorn, M., Lucas, A., Deanfield, J.: Influence of leptin on arterial distensibility: a novel link between obesity and cardiovascular disease? Circulation, 15:1919-1924 (2002).

Singh-Manoux, A., Adler, N., Marmot, M.: Subjective social status: its determinants and its association with measures of ill-health in the Whitehall II study, Social Science and Medicine, 56:1321-1333 (2003).

Skierski, M., Grundland, A., Tuszynski, J.: Analysis of the three-dimensional time-dependent Landau-Ginzburg equation and its solutions. Journal of Physics A (Math. Gen.) 22:3789-3808 (1989).

Smith, G., Hart, C., Blane, D., Hole, D.: Adverse socioeconomic conditions in childhood and cause-specific adult mortality: prospective observational study, British Medical Journal, 317:1631-1635 (1998).

Smith, T., Ruiz, J.: Psychosocial influences on the development and course of coronary heart disease: current status and implications for research and practice, Journal of Consulting and Clinical Psychology, 70:548-568 (2002).

Snow E.: The role of DNA repair in development, Reproductive Toxicology, 11:353-365 (1997).

Somers, S., Guillou, P.: Tumor strategies for escaping immune control: implications for psychoimmunotherapy", in Lewis C., O'Sullivan C. and Barraclough J. (eds.), The Psychoimmunology of Cancer: Mind and body in the fight for survival, Oxford Medical Publishing, pp. 385-416, (1994).

Spiegel, K., Leproult, R., Van Cauter, E.: Impact of sleep debt on metabolic and endocrine function, The Lancet, 354:1435-1439 (1999).

Sternberg E.: Neuroendocrine regulation of autoimmune/inflammatory disease, Journal of Endocrinology 169:429-435 (2001).

Strauss, R., Pollack H.: Epidemic increase in childhood overweight, 1986-1998, Journal of the American Medical Association, 286:2845-2848 (2001).

Svanes C., Omenaas E., Heuch J., Irgens L., Gulsvik A.: Birth characteristics and asthma symptoms in young adults: results from a population-based cohort study in Norway, European Respiratory Journal, 12:1366-1370 (1998).

Tauber, A.: Conceptual shifts in immunology: Comments on the 'two-way paradigm'. In K. Schaffner and T. Starzl (eds.), Paradigm Changes in Organ Transplantation, Theoretical Medicine and Bioethics, 19:457-473 (1998).

Tenallion, O., Taddei, F., Radman, M., Matic, I.: Second-order selection in bacterial evolution: selection acting on mutation and recombination rates in the course of adaptation, Research in Microbiology, 115:11-16 (2001).

Texeira J., Fisk N., Glover V.: Association between maternal anxiety in pregnancy and increased uterine artery resistance index: cohort based study, British Medical Journal, 318:153-157 (1999).

Thaler, D.: Hereditary stability and variation in evolution and development, Evolution and Development, 1:113-122 (1999).

Thayer, J., Lane, R.: A model of neurovisceral integration in emotion regulation and dysregulation, Journal of Affective Disorders, 61:201-216 (2000).

Thayer, J., Friedman, B.: Stop that! Inhibition, sensitization, and their neurovisceral concomitants, Scandinavian Journal of Psychology, 43:123-130 (2002).

Thomas, S., Quinn, S.: The Tuskegee Syphilis Study, 1932 to 1972: implications for HIV education and AIDS risk education programs in the black community. American Journal of Public Health, 81:1498-1505 (1991).

Timberlake, W.: Behavior systems, associationism, and Pavolvian conditioning. Psychonomic Bulletin, Rev. 1:405-420 (1994).

Toulouse, G., Dehaene, S., Changeux, J.: Spin glass model of learning by selection. Proceedings of the National Academy of Sciences 83:1695-1698 (1986).

Turner, B.: Histone acetylation and an epigeneticv code. Bioessays 22:836-845 (2000).

Ullmann, J.: The Anatomy of Industrial Decline, Greenwood-Quorum Books, Westport, CT (1998).

Vallee, R.: Perception, decision and action, Journal of Biological Systems, 2:43-53 (1994).

Volanakis, J.: Human C-reactive protein: expression, structure, and function, Molecular Immunology, 38:189-197 (2001).

Waddington, C.: Epilogue, in C. Waddington (ed.), Towards a Theoretical Biology: Essays, Aldine-Atherton, Chicago (1972).

Waliszewski, P., Molski, M., Konarski, J.: On the holistic approach in cellular and cancer biology: nonlinearity, complexity, and quasi-determinism of the dynamic cellular network, Journal of Surgical Oncology, 68:70-78 (1998).

Wallace, D.: The resurgence of tuberculosis in New York City: a mixed hierarchically and spatially diffused epidemic. American Journal of Public Health, 84:1000-1002 (1994).

Wallace, D.: Analytical approaches for public health from ecosystem science. Proceedings of the 1995 Public Health Conference on Records and Statistics: Data Neees in an Era of Health Reform. CDC/NCHS, Bethesda (1995).

Wallace, D., Wallace, R.: A Plague on Your Houses, Verso Publications, New York (1998).

Wallace D, Wallace R.: Geography of asthma and diabetes over eight US metropolitan regions. Journal of Environmental Disease and Health Care Planning, 3:73-88 (1998).

Wallace, D., Wallace, R.: Scales of geography, time, and population: the study of violence as a public health problem. American Journal of Public Health, 88:1853-1858 (1998).

Wallace, D., Wallace, R.: Life and death in Upper Manhattan and the Bronx: toward an evolutionary perspective on catastrophic social change, Environment and Planning A, 32:1245-1266 (2000).

Wallace, D., Wallace, R., Rauh, V.: Community stress, demoralization, and body mass index: evidence for social signal transduction. Social Science and Medicine, 56:2467-2478 (2003).

Wallace, R., Wallace, D.: Origins of public health collapse in New York City: the dynamics Of planned shrinkage, contagious urban decay, and social disintegration. Bulletin of the New York Academy of Medicine, 66:391-434 (1990).

Wallace R., Fullilove M., Flisher, A.: AIDS, violence and behavioral coding: information theory, risk behavior, and dynamic process on core-group sociogeographic networks, Social Science and Medicine, 43:339-352 (1996).

Wallace R., Wallace D.: Community marginalization and the diffusion of disease and disorder in the United States, British Medical Journal, 314:1341-1345 (1997).

Wallace, R., Wallace, D., Andrews H.: AIDS, tuberculosis, violent crime and low birthweight in eight US metropolitan areas: stochastic resonance and the diffusion of inner-city markers, Environment and Planning A, 29:525-555 (1997).

Wallace, R., Wallace, R.G.: Information theory, scaling laws and the thermodynamics of evolution, Journal of Theoretical Biology, 192:545-559 (1998).

Wallace, R., Wallace, R.G.: Organisms, organizations, and interactions: an information theory approach to biocultural evolution, BioSystems, 51:101-119 (1999).

Wallace, R., Fullilove, R.: Why simple regression models work so well describing risk behaviors in the USA, Environment and Planning A, 31:719-734 (1999).

Wallace, R., Wallace, D., Ullmann, J., Andrews, H.: Deindustrialization, inner-city decay, and the diffusion of AIDS in the USA, Environment and Planning A, 31:113-139, (1999).

Wallace, R., Wallace, D.: Emerging infections and nested martingales: the entrainment of affluent populations into the disease ecology of marginalization, Environment and Planning A, 31:1787-1803 (1999).

Wallace, R., Wallace, R.G.: Immune cognition and vaccine strategy: beyond genomics, Microbes and Infection, 4:521-527 (2002).

Wallace, R., Wallace, D., Wallace, R.G.: Toward cultural oncology: the evolutionary information dynamics of cancer, Open Systems and Information Dynamics, 10:159-181 (2003a).

Wallace, R., Wallace, D., Wallace, R.G.: Coronary heart disease, chronic inflammation, and pathogenic social hierarchy: a biological limit to possible reductions in morbidity and mortality, Journal of the National Medical Association, 96:609-619 (2003b).

Wallace, R., Wallace, D.: Structured psychosocial stress and the US obesity epidemic, Journal of Biological Systems, 13:363-384 (2005).

Wallace, R.: Language and coherent neural amplification in hierarchical systems: renormalization and the dual information source of a generalized stochastic resonance, International Journal of Bifurcation and Chaos, 10:493-502 (2000).

Wallace, R.: Immune cognition and vaccine strategy: pathogenic challenge and ecological resilience, Open Systems and Information Dynamics, 9:51-83 (2002a).

Wallace, R.: Adaptation, punctuation and rate distortion: non-cognitive 'learning plateaus' in evolutionary process, Acta Biotheoretica, 50:101-116 (2002b).

Wallace, R.: Systemic lupus erythematosus in African-American women: cognitive physiological modules, autoimmune disease, and pathogenic social hierarchy, Advances in Complex Systems, 6:599-629 (2003a).

Wallace, R., Wallace, R.G.: Adaptive chronic infection, structured psychosocial stress, and medical magic bullets, BioSystems, 77:93-108 (2004).

Wallace, R., Wallace, D.: 2005, Structured psychosocial stress and the US obesity epidemic, Journal of Biological Systems, 13:363-384 (2005).

Wallace, R.: Comorbidity in psychiatric and chronic physical disease: autocognitive developmental disorders of structured psychosocial stress, Acta Biotheoretica, 52:71-93 (2004).

Wallace, R.: Consciousness: A mathematical treatment of the global neuronal workspace model, Springer-Verlag, New York (2005a).

Wallace, R.: A global workspace perspective on mental disorders, Theoretical Biology and Medical Modelling, 2/1/49 (2005b).

Wallace, R.: Machine hyperconsciousness.
Available at http://cogprints.org/5005/ (2006).

Wallace, R.: Culture and inattentional blindness. Journal of Theoretical Biology 245, 378-390 (2007).

Wallace, R.: Toward formal models of biologically inspired, highly parallel machine cognition. International Journal of Parallel, Emergent, and Distributed Systems 23:367-408 (2008).

Wallace, R.: Developmental disorders as pathological resilience domains. Ecology and Society, 13: 29
http://www.ecologyandsociety.org/vol13/iss1/art29/ (2008).

Wallace, R.: Programming coevolutionary machines: the emerging conundrum. In press, International Journal of Parallel, Emergent, and Distributed Systems (2009).

Wallace, R., Wallace, D., Fullilove, M.: Community lynching and the US asthma epidemic, http://cogprints.ecs.soton.ac.uk (2003).

Wallace, R., Fullilove, M.: Collective Consciousness and its Discontents: Institutional Distributed Cognition, Racial Policy, and Public Health in the United States. Springer, New York (2008).

Wallace, R., Wallace, D.: Punctuated equilibrium in statistical models of generalized coevolutionary resilience: how sudden ecosystem transitions can entrain both phenotype expression and Darwinian selection. Transactions on Computational Systems Biology IX LNBI 5121, 23-85 (2008).

Wallace, R., Wallace, R.G.: On the spectrum of prebiotic chemical systems: an information-theoretic treatment of Eigen's paradox, Origins of Life and Evolution of Biospheres, 38:419-455 (2008).

Wallace, R.G.: AIDS in the HAART era: New Yorks heterogeneous geography. Social Science and Medicine, 56:1155-1171 (2003).

Wallace, R.G., and Wallace, R.: Evolutionary radiation and the spectrum of consciousness. Consciousness and Cognition 18:160-167 (2009).

Waterland R., Michels, K.: Epigenetic epidemiology of the developmental origins hypothesis. Annual Reviews of Nutrition 27:363-388 (2007).

Weaver, I.: Epigenetic effects of glucocorticoids. Seminars in Fetal and Neonatal Medicine, doi:10.1016/j.siny.2008.12.002 (2009).

Weinstein, A.: Groupoids: unifying internal and external symmetry. Notices of the American Mathematical Association 43:744-752 (1996).

Wellman, N., Friedberg, B.: Causes and consequences of adult obesity: health, social and economic impacts in the United States, Asia Pacific Journal of Clinical Nutrition, Suppl.8:S705-S709, (2002).

West-Eberhard, M.: Developmental plasticity and the origin of species differences. Proceedings of the National Academy of Sciences 102:6543-6549 (2005).

Whitehead, M., Judge, K., Bensaval, M., Shouls, S., Diderichsen, F., Hort, S., Danielsson, M.: Project Details: Health Variations Program. Award H28251029.

http://www.lancs.ac.uk/fss/apsocsci/hvp/projects/whitehead.htm (1997-1998).

Whittingham, S., Mackey, I.: The 'pemphigus' antibody and immunopathologies affecting the thymus, British Journal of Dermatology 84:1-6 (1971).

Wichen, E.: Tuberculosis in the Negro in Pittsburgh. Tuberculosis League of Pittsburgh (1934).

Wiegand, R.: An analysis of cooperative coevolutionary algorithms. PhD Thesis, George Mason University (2003).

Wilken, J., Smith, B., Tola K., Mann, M.: Trait anxiety and prior exposure to non-stressful stimuli: effects on psychophysiological arousal and anxiety, International Journal of Psychophysiology, 37:233-242 (2000).

Wilkinson, R.: Unhealthy Societies: the afflictions of inequality, Routledge, London and New York (1996).

Williams, D., Collins, C.: Racial residential segregation: a fundamental cause of racial disparities in health. Public Health Reports, 116:404-416 (2001).

Wilson, K.: Renormalization group and critical phenomena. I Renormalization group and the Kadanoff scaling picture, Physical Revies B 4:3174-3183 (1971).

Wilson, J., Nugent, N., Baltes, J., Tokunaga, S., Canic, T., Yourn, B., Bellinger, E., Delac, D., Golston G., Henderson, D.: Effects of low doses of caffeine on aggressive behavior of male rats, Psychological Reports, 86:941-946 (2000).

Wolfe WS, Campbell CC, Frongillo EA Jr, Haas JD, Melnik TA. 1994. Overweight schoolchildren in New York State: prevalence and characteristics. Am J Public Health, 84:807-813.

Wolpert, D., Macready, W.: No free lunch theorems for search, Santa Fe Institute, SFI-TR-02-010 (1995).

Wolpert, D., Macready, W.: No free lunch theorems for optimization, IEEE Transactions on Evolutionary Computation 1:67-82 (1997).

Wright, R., Rodriguez, M., Cohen, S.: Review of psychosocial stress and asthma: an integrated biopsychosocial approach, Thorax, 53:1066-1074 (1998).

Wymer, C.,: Structural nonlinear continuous-time models in econometrics. Macroeconomic Dynamics 1:518-548 (1997).

Yammamoto, A., Ishihara, K.: Penrose patterns and related structures. II. Decagonal quasicrystals, Acta Crystallographa A, A44:707-714 (1988).

Yerger, V., Malone, R.: African American leadership groups: smoking with the enemy. Tobacco Control, 11:336-345 (2002).

Zahner, G., Kasl, S., White, M., Will, J.: Psychological consequences of infestation of the dwelling unit. American Journal of Public Health, 75:1303-1307 (1985).

Zhu, R., Rebirio, A., Salahub, D., Kaufmann, S.: Studying genetic regulatory networks at the molecular level: delayed reaction stochastic models, Journal of Theoretical Biology 246, 725-745 (2007).

Index

acetylation, 3
adiabatic, 32
adipocytes, 17
algorithmically random, 33
apoptosis, 10
APSE, 32
asthma, 7, 22, 138
Atmanspacher H., 69
autoimmune disease, 136
autoimmune disorder, 121

Baars B., 28, 70
Bailey N., 23
Bennett C., 35
Bird A., 3
Bjorntorp P., 18
blood pressure, 10
BMI, 16, 147
Boyd R., 7
breast cancer, 7

cancer, 111
central abdominal obesity, 22
chain rule, 45
chromatin, 2
chronic inflammation, 99, 106, 119
Ciliberti S., 27
circadian cycles, 127
cirrhosis deaths, 147
coevolutionary development, 53
coevolutionary programming, 67
cognition as language, 12
cognitive paradigm, 8
Cohen I., 1, 8

comorbid, 7
comorbidity, 14
consciousness, 11
contagious risk behavior, 19
coronary heart disease, 7, 99
cortisol, 17
Crews D., 4
Crick F., 70
critical period, 63, 64
CUES, 161
cultural lock in, V, 71, 181
culture, 71

D-sugar, 43
Darwinian individuality, 43
deindustrialization, 82
deurbanization, 82
developmental disorders, 59
diabetes, 7
diabetes deaths, 89
directed homotopy, 29, 65
disease guild, 178, 181
dose-response relation, 93
drug deaths, 147
Durham W., 7

E-property, 33
ecological control program, 97
ecological domain shift, 89
ecological resilience, 63
ecosystem as information source, 60
ecosystem resilience, 6
Eldredge N., 7
emotion, 11

endocrine disruption, 154
engineering resilience, 6
epigenetic catalysis, 45, 67, 71, 124, 142, 181
epigenetic epidemiology, 2
epigenetic focus, 56
epigenetic programming, 66
epigenetic regulation, 45
equivalence class algebra, 37
ergodic, 32

farmer, 60
Feynman R., 2, 35
Foley D., 4
free energy density, 13

genon, 6
Glazebrook J., 2, 41
glucocorticoids, 5
Gould S., 7
Granovetter, M., 157
groupoid, 37, 193
groupoid free energy, 39
groupoids, 29
Guerrero-Bosagna C., 2
Gunderson L., 7, 63

health disparities, 73
Hirsch J., 81
histones, 3
HIV/AIDS, 161
Holling C., 7, 20
HPA axis, 5, 10, 74, 133
Human Genome Project, 1
hypertension, 7

ICCS, 147
immune system, 8
inattentional blindness, 57, 61
inverted U, 157
isotropy group, 194

Jablonka E., 2
Jaenish R., 3
Joint Asymptotic Equipartition Theorem, 45
Jost J., 6

L-forms, 43
Lamb M., 2

Landau L., 40
leptin, 15, 17, 100
Link, B., 162
lupus, 121

MacReady W., 56
mass unemployment, 85
material deprivation, 163
Maturana H., 1
mereological fallacy, 69
mesoscale resonance, 142
metabolic syndrome, 22
methylation, 2
Miller/Urey experiment, 41
Mokdad A., 16
Morse Function, 51
Morse Theory, 196
multiple processes, 52
mutator, 114

neighborhood conditions, 22
network information theory, 59
neural network models, 28
New York Academy of Medicine, 161
NHANES, 16
Nix/Vose models, 67
no free lunch theorem, 56
nonorthogonal eigenmodes, 124
Nunney L., 10
NYC fires, 82

O'Nuallain S., 1
obesity, 7, 15, 73
occupational grade, 17
Onsager relations, 198

partition function, 38, 49
Phelen, J., 162
phosphorylation, 3
prostate cancer, 7
Ptolemaic Theory, 69

racemic, 41
rate distortion dynamics, 47
redlining, 20
resilience topologies, 12
Richerson P., 7
rust belt, 85

Scherrer K., 6

Shannon Coding Theorem, 183
Shannon-McMillan Theorem, 33
Siefert-Van Kampen Theorem, 29
signal transduction, 148
simply connected, 29
socioeconomic disintegration, 82
species guild, 173
spinglass model, 27, 38
spontaneous symmetry breaking, 40, 50
stationary, 32
stochastic resonance, 148
sumoylation, 3

topological hypothesis, 40
Toulouse G., 28
transcriptional regulators, 27

tumorigenesis, 10, 113
tunable epigenetic catalysis theorem, 67
tuning theorem, 185
Turner B., 3

ubiquitination, 3
Upper Manhattan, 143

Varela F., 1
viable networks, 33

Weaver I., 5
West-Eberhard M., 4, 46
Whitehall Studies, 18, 74
Wolpert D., 56

Breinigsville, PA USA
30 April 2010
237133BV00005B/48/P